T0137100

Studies in Big Data

Volume 54

Series editor

Janusz Kacprzyk, Polish Academy of Sciences, Warsaw, Poland
e-mail: kacprzyk@ibspan.waw.pl

The series "Studies in Big Data" (SBD) publishes new developments and advances in the various areas of Big Data—quickly and with a high quality. The intent is to cover the theory, research, development, and applications of Big Data, as embedded in the fields of engineering, computer science, physics, economics and life sciences. The books of the series refer to the analysis and understanding of large, complex, and/or distributed data sets generated from recent digital sources coming from sensors or other physical instruments as well as simulations, crowd sourcing, social networks or other internet transactions, such as emails or video click streams and others. The series contains monographs, lecture notes and edited volumes in Big Data spanning the areas of computational intelligence including neural networks, evolutionary computation, soft computing, fuzzy systems, as well as artificial intelligence, data mining, modern statistics and operations research, as well as self-organizing systems. Of particular value to both the contributors and the readership are the short publication timeframe and the world-wide distribution, which enable both wide and rapid dissemination of research output.

** Indexing: The books of this series are submitted to ISI Web of Science, DBLP, Ulrichs, MathSciNet, Current Mathematical Publications, Mathematical Reviews, Zentralblatt Math: MetaPress and Springerlink.

More information about this series at http://www.springer.com/series/11970

Moses Eterigho Emetere

Environmental Modeling Using Satellite Imaging and Dataset Re-processing

 Springer

Moses Eterigho Emetere
Department of Physics
Covenant University
Ota, Nigeria

ISSN 2197-6503 ISSN 2197-6511 (electronic)
Studies in Big Data
ISBN 978-3-030-13407-5 ISBN 978-3-030-13405-1 (eBook)
https://doi.org/10.1007/978-3-030-13405-1

Library of Congress Control Number: 2019931821

This Springer imprint is published by the registered company Springer Nature Switzerland AG
The registered company address is: Gewerbestrasse 11, 6330 Cham, Switzerland

This work is dedicated to the memory of my father—Christopher Ukoko Emetere

Preface

Despite the awareness of the "Big data" concept, more than 90% of information are discarded due to limited storage system and incompetent analytic techniques. Till date, only few information can be extracted from large imaging dataset. This challenge is considered to be very pertinent in environmental modeling. This book illustrates unique techniques for re-processing images to extract numerical information that can be used to quantify observables. Hence, experiments or procedures that yield large images can be statistically or parametrically examined—through the use of open-source libraries. New techniques for re-processing image and manage "big data" was propounded using open-source libraries/packages. Therefore, "big data" in the form of images or ASCII dataset can be comparatively analyzed along same defined procedures or standards.

The purpose of writing this book is to solve the challenges of discarding dataset or images which are relevant directly or indirectly to the research. The procedure of dataset screening leads to the discard of over 90% of the original dataset or images generated in experiments or survey. If the images or dataset are generated under same principles or conditions, then each measurement maybe a narrative of the actual scenario. The dataset that was used for illustration purpose was obtained from satellite sources. The types of dataset were satellite images of aerosol optical depth (AOD) at 500 and 550 nm, aerosol optical depth pixel count (AODPC), dust aerosol column optical depth (DACOD) at 550 nm, and aerosol absorption optical depth (AAOD) at 388 nm. A large dataset pull in the form of images or ASCII dataset for AOD (at 440, 555, 670, and 865 nm) over 170 locations were analyzed. Thirteen years' dataset was considered for each of the locations. The statistical analysis of satellite images was illustrated for studying large volume of images. The modalities for planning a computational or mathematical model and its corresponding results were highlighted and interpreted.

This book solves the problem of discarding measurements based on its volume and unquantifiable nature of the dataset. This book is recommended for research assistants, professionals in the fields of Environmental Sciences, Civil Engineering, Chemical Engineering, Environmental Chemistry, Atmospheric Physics, Space Physics, Environmental Biology, and Communication/Electrical Engineering.

Also, this book is specifically recommended for research interns and scientists in communication and satellite industries; research interns and scientists in environmental monitoring centers/institutes; and research assistants working on "Image processing" and "big data" in environmental studies.

Ota, Nigeria   Moses Eterigho Emetere

Contents

Chapter 1
Introduction to Environmental Modelling

Many scholars have defined environmental modelling according to their disciplinary inclination. For example, DBW (2018) defines environmental modelling as "the application of multidisciplinary knowledge to explain, explore and predict the Earth's response to environmental change, both natural and human-induced". Wiki (2018) defined environmental modelling as "the creation and use of mathematical models of the environment. It is generally done either for pure research purposes, or inform decision making and policy". FES (2018) defined environmental modelling as the "modelling of natural processes associated with inanimate nature, such as hydrological and hydraulic modelling, modelling of chemical processes and processes in the atmosphere". Hauduc et al. (2015) defines environmental modelling as the process of accounting "for multiple variables and multiple objectives in systems with many processes occurring at different time scales". Delft3D (2018) define environmental modelling as the model that "is based on the transport of substances using the so-called advection-diffusion equation". FA (2018) defines environmental modeling as a model that "focus on only noise or emission outputs that presents a need to better consider noise, air quality, fuel burn, and greenhouse gas emissions interdependencies and their costs and benefits". EVO (2018) define environmental modelling as a model "that enable raw data to be transformed into useful information, through synthesis, simulation and prediction". Wiki (2018) defines environmental modelling as the "the process of using computer algorithms to predict the distribution of species in geographic space on the basis of a mathematical representation of their known distribution in environmental space".

In my own words, environmental modelling is building or developing efficient working systems (which may be in form of computational or mathematical or statistical or spatial application) to estimate, evaluate or mimic a real environmental situation with the aim to showing adequate understanding of the concept; manipulating or optimizing known parameters; and proffering sound solution(s) that can assist decision making process(es). There are different types of environmental models mentioned in literature. Figure 1.1 describes the pictorial concept of environmental modelling.

© Springer Nature Switzerland AG 2019
M. E. Emetere, *Environmental Modeling Using Satellite Imaging and Dataset Re-processing*, Studies in Big Data 54,
https://doi.org/10.1007/978-3-030-13405-1_1

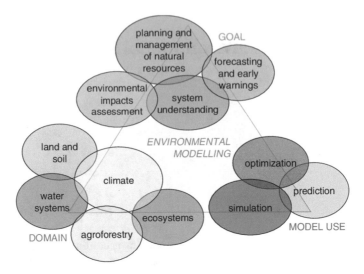

Fig. 1.1 Pictorial description of environmental modelling (Ogola 2007)

In the concept of Ogola (2007) as shown above, environmental modelling is divided into three subgroups, namely field application, known outcome and processing techniques. The 'field application' is made-up of water systems, land and soil, climate, ecosystems and agroforestry. In addition to the 'field application', there exist the renewable energy systems, geo-disciplines and atmospheric systems. Atmospheric systems are almost synonymous to the climate system, however, the distinction between these two systems is that atmospheric systems looks at the environmental forces that initiates wind systems, rainfall, surface temperature, snow cover etc. According to Ogola (2007), the 'known outcome' includes planning and management of natural resources, system understanding, forecasting and early warning, environmental impacts assessment. Lastly, the processing techniques include simulation, optimization and prediction. In addition to the concept of processing techniques is model development, model verification and model validation. A typical model development technique is presented by Logica (2018) in Fig. 1.2.

The quantitative environmental model has been described as a model that focuses on research, management and decision-making. The Integrated environmental model is the integration of multiple modeling techniques from different disciplines to solve environmental problems. Computational environmental model is the use of software applications for predictive or goal-oriented studies for specific environmental problems. Spatial environmental model describes an analytical process used to estimate or evaluate properties of spatial features that are gotten from geographical information system (GIS) or satellite imagery.

Researcher or modelers' routine in model formulation or application includes model calibration, validation, verification and sensitivity analysis. Model calibration is the process of assigning values to parameters, terms and constants. These values

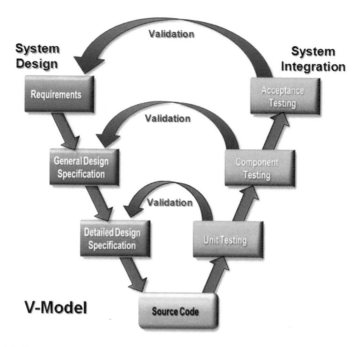

Fig. 1.2 Model development techniques (Logica 2018)

are used as input in the model to produce numerical output. Model validation is a process used for showing how the new model meets-up with some known standard. This standard may be the numerical output or behavioural trend of an existing work. Model verification is a process of proving that the modeling formalism is correct. This process includes debugging each compartment of the computer program; testing the mathematical models with live dataset; showing that the program logic is correct; comparing the sensitivity of the model with an existing model; re-scaling to detect errors or uncertainties in the model etc. Sensitivity analysis is a process which the modeler evaluates the responses of the model by considering changes in input parameters. Most time, the modeler or researcher develops standards to show that his/her model could assimilate dataset and comprehend every single detail or modification to the input parameters. In this chapter, the general outline on environmental models and few applications were discussed. Lastly, few aerosols model were considered.

1.1 General Outline of Environmental Models

In the general sense of environment model, its scope is quite bogus considering the multi-disciplinary interpretation of the word 'environment'. In the field of economics, environmental model is referred to as 'environmental-economic models' that describes a qualitative or quantitative way of identifying least-cost policies or policy mixes of reducing environmental hazards. For example, Lanzi (2017) and OECD (2014) gave clear details on economic consequences of air pollution. Also, the economic consequences of climate model have been discussed by OECD (2015). In the field of psychology, environment model relates the behavior that emanates from changing environment and how that environment affects its inhabitants. Xiang et al. (2017) conducted a research on psychological and behavioral effects of air pollution, and how these effects are developed under different theoretical framework. It was argued that psychosomatic status was also responsible for adverse effect of air pollution.

From the above, it is quite interesting to note that environmental model is a multi-disciplinary affair and cannot be hinged mainly on fields of science and engineering. Environmental modelling can be divided into five broad types i.e. hydrology, climate, ecological, soil/geological and psychology/economy. The hydrological environmental models include surface water models, surface water runoffs model, subsurface water models and coastal models. From literature, surface water model can be represented mathematically in form of one and two dimensional models. The one-dimensional surface water models show a derivation that has multiple cross - sections perpendicular to the anticipated flow path (Zhang et al. 2013; Strong and Zundel 2014). Also, the one dimensional surface water model is characterized by the use of step-backwater methodology to determine water surface elevation and average flow velocity within each cross-section. The two dimensional surface water model considers the discretization of study location into grids; the determination of its water surface elevation within each grid element; and the estimation of flow into each adjacent grid element using finite difference method (Yoshioka et al. 2014). However, some researchers have adopted other methods for solving the two dimensional surface water model (Bai et al. 2016).

The difference between the one and two dimensional models can be summarized in the complexity of resolving parameters like flow time, grid elements, design configurations of the model, topography of the study area etc. Hence, the complexity of the surface water model does not depend on its features (i.e. one or two or three dimensional model) but on the researcher perception of the model design. The surface water runoffs model is a sub-division of the surface water model. Bhatt and Mall (2015) related the surface water runoffs to climate change in a simulation model. Like the surface water models, the subsurface water models are discussed with respect to its dimensions i.e. 1D, 2D or 3D. One of the famous subsurface water models is the Hydrus model. The Hydrus model has one (Hydrus-1D), two (Hydrus-2D) and three dimensional (Hydrus-3D) models. Hydrus 3D model theoretical frameworks is derived from the Richard's equation.

The coastal model relates the influence of the marine environment on terrestrial environment and vice versa. This concept is related to the sea-level rise. The dynamics of sea-level changes is crucial because it has great influence on the terrestrial environment. Church et al. (2001) propounded that human-induced global warming is a major cause of the global-mean sea-level rise that leads to an increase in the global volume of the ocean. Emetere and Akinyemi (2018) reported a computational environmental model that accurately evaluated the implication of climate change on the sea level changes in seven stations on the upper Atlantic.

The ecological environmental model is the second largest concept in environmental research. The sub-division of the ecological model cuts-across the specialized branches of ecology (Samiksha 2017) namely: habitat ecology, community ecology, population ecology, evolutionary ecology, taxonomic ecology, human ecology, applied ecology, ecosystem dynamics, ecological energetic, ecophysiology, genecology, paleoecology, ecogeography, pedology, ethology etc. For example, paleoecology is the study of the life of the past ages through the instrumentality of proven methods as palynology, paleontology, and radioactive dating. Seddon et al. (2013) summarized past paleoecology study and models in fifty salient perspectives.

The climate environmental model is adjudged the broadest research in environmental studies because it integrates broad topics—ocean, atmosphere, land surface, space, solar system etc. Among the specialized branches of climate model, the atmospheric researches have attracted more scientific publications in the past five decades. The quest to explain climate change and its numerous effects on the environment has increased research prospects in atmospheric studies or research. Also, atmospheric research is embedded in other specialized branches of climate modelling (Emetere 2014, 2016a, 2017a, b, c; Emetere et al. 2016; Emetere and Akinyemi 2017).

The soil/geological model is a compilation of all interactions below the soil that affects the terrestrial environment. For example, earthquakes, land-slides, earth tremor etc. are very vital in environmental modelling. Researchers have shown that models can be propounded to monitor, estimate and evaluate events below the earth surface (Gupta and Jangid 2011; Emetere 2017d). Also, some researchers have also modelled the effects of geological disturbances on the eco-system and atmosphere (Devine et al. 1984; Akinyemi et al. 2016).

There are many sub-modeling techniques in environmental modelling that help to understand certain phenomenon. For example, conventional modeling deals with the process of creating a simplified representation of reality to understand it and potentially predict and control its future development. Edward et al. (2009) used the conventional method to investigate carbon dioxide emission in trucks and vehicle. It was observed that typical van-based vehicle produced 181 g of carbon dioxide (CO_2), compared with 4274 g of CO_2 for an average trip by car and 1265 g of CO_2 for an average bus passenger. Jinduan and Dominic (2018) used the conventional modelling to investigate the short term water demand using daily water demand, daily maximum air temperature, and daily total rainfall data from Lexington, Ky., to develop and test several forecast models. Ghumman et al. (2011) used the conventional model to forecast rainfall run-off in the watershed in Pakistan. It was observed that conventional model maybe considered as an important alternative to conceptual models and it can be used when the range of collected dataset is short and of low standard.

Integrated modeling is a sub-modelling technique in environmental modelling and it refers to combination of a set of interdependent science-based components (models, data, and assessment methods) to form an appropriate modeling system. Hughes et al. (2011) used the integrated model to solve ground water problem in Thames.

The two models have been connected using the model linking standard OpenMI. The OpenMI (Open Modelling Interface) is an open IT standard that facilitates linking hydrological model with modules. The primary objectives of OpenMI platform are to develop, maintain and promote the use of multiple models. Johnston et al. (2011) used the integrated model to predict the state of freshwater ecosystem services within and across the Albemarle-Pamlico Watershed, North Carolina and Virginia (USA). The integrated model is made up of five environmental models that are linked within the Framework to provide multimedia simulation capabilities. The models are: the Soil Water Assessment Tool (that predicts watershed runoff); the Watershed Mercury Model (that simulates mercury runoff and loading to streams); the Water quality Analysis and Simulation Program (that predicts water quality within the stream channel); the Habitat Suitability Index model (that predicts physicochemical habitat quality for individual fish species); and the Bioaccumulation and Aquatic System Simulator (that predicts fish growth and production, as well as exposure and bioaccumulation of toxic substances).

Integrated assessment modeling is a sub-modeling technique in environmental modelling that considers an analytical approach to integrate knowledge from a variety of disciplinary sources to describe the cause-effect relationships by studying the relevant interactions and cross-linkages. Rotmans and van Asselt (2001) used the integrated assessment modelling to examine the history, general features, classes of models, strengths and weaknesses, and the dilemmas or challenges researcher encounter. However, there are uncertainties associated to the integrated model. This includes erroneous knowledge or data, inherent variability (Cullen and Frey 1999). Matott et al. (2009) reported that the total uncertainty of a given quantity may be characterized in one of four ways: purely irreducible (i.e. the quantity varies and the associated population has been completely sampled without error); partly reducible and partly irreducible (i.e. the quantity varies and the associated population has been partially sampled or sampled with error); purely reducible (i.e. the quantity does not vary but has been sampled with error); and certain (i.e. the quantity does not vary and has been sampled without error).

Probabilistic model (statistical or stochastic models) is referred as a sub- modelling technique that utilize the entire range of input data to develop a probability distribution of model output rather than a single point value. Sun et al. (2014) used the probabilistic model to predict the flow of four engineered nanomaterials (nano-TiO_2, nano-ZnO, nano-Ag and CNT) to the environment and to quantify their amounts in (temporary) sinks such as the in-use stock and ("final") environmental sinks such as soil and sediment.

The model life-cycle is defined as one of the key concepts of systems engineering that generally consists of a series of stages regulated by a set of management decisions to estimate the maturity of the system to transcend from one stage to another. A pictorial definition of model life-cycle is shown in Fig. 1.3.

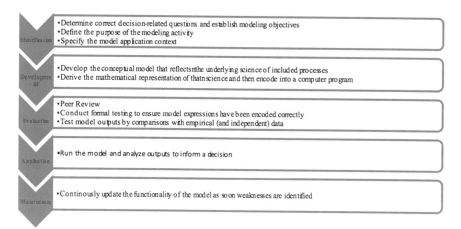

Fig. 1.3 Pictorial overview of model life-cycle

1.2 Dimensions of Environmental Models

Hämäläinen (2015) introduced the concept of behavioural issues in environmental modelling. This includes the salient issues that prompts a modeler or researcher to make decision on: the choice of method for his/her model; the choice of model to examine; the choice of approach in solving problems; the choice of whom to cite; the choice of whom to critic; the choice of research modalities; the mode of explaining model; the choice of whom to collaborate with. As much as it is good to conceive a model, it is essentially paramount for researcher to consider how to develop a model, test-run a model, validate a model, and expand model application. It is also good to understand the reality of the risk of making choices in environmental modeling. For example, Montibeller and von Winterfeldt (2015) highlighted a list of cognitive and motivational biases in decision making. Cognitive biases are systematic patterns of deviating from norm or rationality in judgment. This includes anchoring certain bias, equalizing bias, gain-loss bias, myopic problem representation, splitting biases, proxy bias, range insensitive bias and scaling. According to Hämäläinen (2015), the main goal of considering behavioral issues in environmental modelling is to improve the understanding of decision processes and to produce better outcomes (like predictions, decisions and policies) to avoid 'Hammer and Nail' syndrome in upcoming modelers. 'Hammer and Nail' syndrome is when a modeler or researcher uses only a single modelling tool to solve all kinds of problem. This challenge may emanate from some scientists who are clamoring for expertise in the use of a single tool. This kind of cognitive bias is called anchoring. In environmental modelling, veracity in understanding many tools is very important as it gives the modeler new perspectives to attain high level accuracy.

Generally, in scientific discuss, the 'Bandwagon effect' bias is very common. This kind of cognitive bias is sometimes referred to 'herds thinking' where there is the tendency to do (or believe) things because many other people do (or believe) the same. The recent trends in environmental modelling are worrisome because 'Bandwagon effect' is massively crumbling the possibility of creating new perspectives in a given field. Some 'old school' researchers (i.e. those who believe strictly on a concept) are forcibly rejecting opposing scientific hypothesis that seem to negate their believes on the validity of certain scientific postulations. This challenge leads to the question—is there a perfect model? Sterman (2002) answer to the question clearly shows that there is no perfect model, however, the usefulness of a model is a function of its perceived relevance.

Another common type of bias environmental researcher face is the 'Bias blind spot'. Victims of this kind of bias see their self as less biased than other people. This challenge has created wide disparity amongst scientists as it has played down on superior facts and upheld superior complexes. The beauty of knowledge is that we all cannot see from the same vintage point. While some observer can effortlessly explain their own side of a matter (may be due to less complexity of the observables), other observer may have difficulty to comprehend the complexities of observables. Some literatures in environmental studies are offensive because the writer castigates so many research works without a substantial evidence to support their claims.

The 'Continued influence effect' is a known cognitive bias in environmental modelling that creates a virtuous cycle for a long period of time. This type of bias occurs when there is the tendency to believe previously learned misinformation even after it has been corrected. A typical example is the application or functionality of the general circulation models (GCMs). Wilby et al. (2002) argued that GCMs are restricted in their usefulness for local impact studies by their coarse spatial resolution (typically of the order 50,000 km^2) and unable to resolve important sub-grid scale features such as clouds and topography. However, Min-Seop and In-Sik (2018) has successfully resolved the sub-grid scale challenge by considering a three-dimensional cloud resolving model simulation to estimate the appropriate ratios of the sub-grid scale vertical transport to the total vertical transport of moist static energy for different horizontal resolutions in the cumulus base mass flux. The identification of an error in an existing model and the perceived solution to the problem are very important in environmental modelling. Hence, beginners in environmental modelling should consciously avoid this pit-fall because it would make the modeler or researcher exhaust much energy to accomplish any meaningful task.

The opposite of the 'Continued influence effect' bias is the 'Irrational escalation'. This bias occurs when people justify increased investment in a decision, based on the cumulative prior investment, despite new evidence suggesting that the decision was probably wrong. This kind of bias is common where there is the 'herds thinking'. Unfortunately, research institutions are more culpable to exhibit this type of bias. Many upcoming modeler sometimes face the challenge to have this bias based on their perception of the name of the research institute.

Most upcoming modeler have the 'Ostrich effect' bias. This bias entails the victim tendency to ignore obvious (negative) facts in the formulation or modification of a model. The antidote to this bias is consulting many literatures before conceiving the parametric concept of the model.

As discussed earlier, the ultimate objective of an environmental model is its relevance or applicability for policy making, software development etc. In recent research work, modeler or researcher are expected to show the application of their model, its reproducibility, relevance, accuracy etc. Some modeler may go an extra mile in expanding the scope of their environmental model to other branches of environmental studies. For example, the thermographic model has been proven to show great success in interpreting meteorological imbalances (Emetere 2014). This effect was applied to explain the thermal distribution during volcanic eruption (Emetere 2017d). Lastly, the knowledge that was gain from the thermal properties sub-surface elements led to the extensive use of the model to detect hydrocarbon entrapment in the earth surface (Emetere et al. 2017a).

The advance stage of model application is the adoption of the model in regulatory development such as setting standards, or enforce regulatory requirements. For example, the Environment Protection Agency (EPA) has adopted AERMOD application software for setting standards for air pollution dispersion from point sources. AERMOD modeling system includes extensive documentation, model code, user's guide, supporting documents, and evaluating databases, all of which are available on the web site of the EPA Support Center for Regulatory Atmospheric Modeling (NRC 2007).

In like manner, SUTRA (saturated–unsaturated transport) and SUTRA^{-1} models are used as standards for evaluating the accuracy of hydrologic and hydrogeochemical processes i.e. movements of pollutants and water (Bobba et al. 2000). SUTRA was developed in 1984 by United States Geological Survey (USGS). It is a three-dimensional groundwater model that simulates solute transport (i.e. salt water) or temperature in a subsurface environment.

The selection of models as standards sometime may not be hinged on only its accuracy. Sometimes, the selection of models is often based on familiarity. Hence, the criteria for universal acceptance of any model may be the versatility of the inventor or modeler to arose wider usage of the model. This may be the reason why institutional based models are promoted than individual models.

1.3 Aerosol Models

The main focus of this book is to discuss the dynamics of environmental modelling with emphasis on re-processing of satellite imageries of atmospheric aerosol distribution. The definition and sub-division of atmospheric aerosol model is still unclear because its concept is very broad—considering the factors that triggers its distribution, dispersion and particulate life-time. Hence, authors define the types of aerosol models according to their research objectives. For example, Shettle and Fenn (1979) listed the types of aerosols model as rural aerosol model, urban aerosol model, maritime aerosol model, tropospheric aerosol model and fog model. In this section, aerosol models will be discussed based on its basic properties.

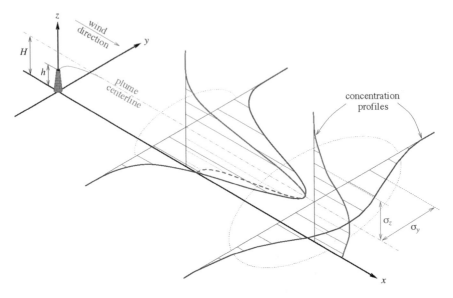

Fig. 1.4 Basics of advection-dispersion models (Stockie 2011)

1.3.1 Advection-Dispersion Models

Advection-dispersion models are mostly developed on regional scale. This is because regional meteorology differs from one geographical location to another. The basics of advection-dispersion models is shown in Fig. 1.4. Till date, it is still a huge task—integrating regional models into a complex global scale. One of the reason is that most regional advection-dispersion models are contested based on its theoretical soundness and computational validity. Zhang et al. (2014) investigated the global atmospheric aerosol transport model using 3D advection-diffusion equations that was an extension of the 2D advection-diffusion equation:

$$\frac{\partial c}{\partial t} + u_x \frac{\partial c}{\partial x} + u_y \frac{\partial c}{\partial y} + u_z \frac{\partial c}{\partial z} = k_x \frac{\partial^2 c}{\partial x^2} + k_y \frac{\partial^2 c}{\partial y^2} + k_z \frac{\partial^2 c}{\partial z^2} + \lambda c \qquad (1.1)$$

where c is contaminant concentration; t is time; u_x, u_y, u_z represent wind speed in the three directions x, y, z respectively; k_x, k_y, k_z represent turbulent diffusivity in three directions; λ is the climatic factor, which can be its emission source, chemical conversion, dry deposition and wet scavenging.

The Euler finite difference method for numerical simulation which has a horizontal resolution $4° \times 5°$ and a vertical direction (divided into 11 sub-layers) was used to resolve Eq. (1.1). The model was queried because the application of the model was based on laboratory framework only.

Fig. 1.5 Basics of box
models (Guwahati 2014)

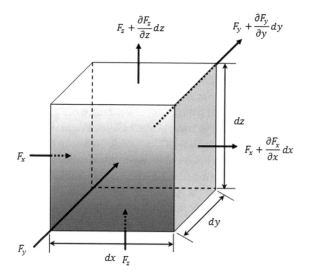

Holmes and Morawska (2006) developed dispersion model whose principles was based on the box model (BM). The BM operates on the principle of conservation of mass. The box model is familiar in atmospheric research. Choo-in (2001) applied the box model to estimate the pollutants in a street tunnel in Thailand. The box model works perfectly when the air mass is well mixed and concentrations are uniform throughout. When the box is not defined, pollutants are formed within the box only; hence, the information on the local concentrations of the pollutants is considered as negligible. The basics of the box-model is shown in Fig. 1.5.

The Gaussian model (GM) is the most popular model used in atmospheric dispersion modeling. Gaussian or plume models operate based on the Gaussian distribution of the 2D or 3D concentration of the plume under steady state conditions. The Gaussian distribution of the plume is under certain influences like turbulent reflection from the surface of the earth, dimension of transport, boundary layer (especially when the mixing height is low), stability classes or travel time from plume sources.

Ahmada (2011) worked on the dispersion of atmospheric pollutants using two dimensional advection diffusion equations. He started with the generalized advection-diffusion equation given below

$$\frac{\partial C}{\partial t} = \mu \frac{\partial^2 C}{\partial x^2} - u \frac{\partial C}{\partial x} \tag{1.2}$$

and obtained the two-dimensional advection-diffusion equation given as

$$\frac{\partial C}{\partial t} - \mu \left(\frac{\partial^2 C}{\partial x^2} + \frac{\partial^2 C}{\partial y^2} \right) + u \frac{\partial C}{\partial x} + v \frac{\partial C}{\partial y} + \varpi C = \frac{f}{H} \tag{1.3}$$

where C is the concentration of pollutants, H is the depth occupied by pollutants, u is the wind velocity or drift velocity, f is the power of the source, ϖ is the pollutant chemical activity coefficient of pollutants and μ is the horizontal diffusion coefficients. The boundary conditions used were the zero Dirichlet boundary condition, Neumann boundary condition and periodic boundary condition. The finite difference approach was adopted in order to obtain the numerical solutions of Eq. (1.3). The solutions for first order forward difference, first order backward difference, first order central difference, second order central difference, central differences for two dimensional functions of the Crank-Nicolson method were determined. The model propounded has three external parameters, namely the pollutant diffusion coefficient μ, the drift velocity of air u and the pollutant chemical activity coefficient ϖ.

Thongmoon et al. (2007) worked on the numerical solution of a 3D advection-dispersion model for pollutant transport-using the forward in time and centre in space (FTCS) finite difference method. The paper is an extension of the Choo-in (2001) box model with a different dimensionality.

$$\frac{\partial C}{\partial t} + u\frac{\partial C}{\partial x} + v\frac{\partial C}{\partial y} = D_h\frac{\partial^2 C}{\partial x^2} + D_h\frac{\partial^2 C}{\partial y^2} + D_v\frac{\partial^2 C}{\partial z^2} \tag{1.4}$$

where u and v are constant wind speeds in the x and y-directions respectively. D_h and D_v are constant dispersion coefficients in the x and z-directions respectively.

Benson et al. (2000) worked on the application of a fractional advection-dispersion equation (FADE). They used fractional derivatives to study the scaling behavior of plumes that undergo Levy motion. This scaling behavior is in time and space of the heavy tailed motion (Daitche and Tamas 2014). However, the second-order dispersion arises for a thin tailed motion. Under this condition, very large motions are completely ruled out. Contrary to the thin tailed motion, the fractional advection-dispersion equation considers a very large transition of particles which arise from high heterogeneity (Benson et al. 2000). The FADE is effective (when the scaled α-stable density is known) to predict distances of particles in closed forms and their concentrations versus time. FADE have been found to be accurate in a laboratory settings. However, its accuracy under geographical uncertainties has not been resolved.

1.3.2 Aerosol Optical Depth: Satellite Retrieval Model

The AOD is a vital parameter that applies to determining air quality that affects: environment and life-forms; monitoring volcanic and biomass pollution; forecasting and now-casting earth radiation budget and climate change; estimating variability of aerosols and its size distribution in the atmosphere. The greater the magnitude of the AOD at specified wavelengths, the lesser the chances of light at that wavelength to reach the Earth's surface. Aerosol optical depth is the measurement of transparency of the atmosphere. When AOD is less than 0.1 and 1.0, it indicates a crystal clear sky and very hazy conditions, respectively. AOD measures the amount of light lost due to the presence of aerosols or aerosols distributed on a vertical path through the atmosphere.

The sun photometer measures the AOD using the Beer-Lambert-Bouguer law where the voltage (V) is directly proportional to the spectral irradiance (I) measured by the sun photometer. The mathematical expression for the Beer-Lambert-Bouguer law (Faccani et al. 2009; He et al. 2012) is given:

$$V(\lambda) = V_o(\lambda)d^2 \exp(-\tau(\lambda)_{tot} \times m) \tag{1.5}$$

where $\tau(\lambda)_{tot}$ is the total optical depth, and m is the optical air mass, V_o is the extraterrestrial voltage, V is the digital voltage measured at wavelength λ, d is the ratio of the average to the actual Earth-Sun distance. The Beer-Lambert-Bouguer equation (He et al. 2012) could also be modified as

$$\tau_a = \frac{\left(In\left(\frac{V_o}{R^3}\right) - In(V - V_{dark}) - a_R \left(\frac{p}{p_o}\right)m \right)}{md} \tag{1.6}$$

where τ_a is the aerosol optical depth, V_o is the calibration constant for the sun photometer, R is the Earth-Sun distance, d is the day of the year, V and V_{dark} are the sunlight and dark voltages from the sun photometer respectively, a_R is the contribution of optical thickness of molecular (Rayleigh) scattering of light in the atmosphere, p is the station pressure, p_o is standard sea level atmospheric pressure, m is the relative air mass and written as $m = \frac{1}{\sin(\theta)}$, θ is the solar elevation angle.

The measurement of AOD is complex because aerosol is not solely responsible for the scattering or absorption of light. Other atmospheric constituents, for example, methane, ozone, nitrogen oxides, carbon (IV) oxide, water vapour scatter or absorb light, hence their joint AOD can be calculated as shown mathematically below (Liu et al. 2011):

$$\tau(\lambda)_{aerosol} = \tau(\lambda)_{tot} - \tau(\lambda)_{water} - \tau(\lambda)_{Rayleigh} - \tau(\lambda)_{O_3} \\ - \tau(\lambda)_{NO_2} - \tau(\lambda)_{CO_2} - \tau(\lambda)_{CH4} \tag{1.7}$$

where $\tau(\lambda)_{Rayleigh}$ is the optical depth of the Rayleigh scattering. Spectral aerosol optical depths at wavelength 440–870 nm are typically used to estimate the size distribution of aerosols. The size distributions of aerosols are better described by the Angstrom parameter (α) which can be calculated using two or more wavelengths. The most popular mathematical representation (Liu et al. 2011) of α is given as

$$\alpha = -\frac{d \, In \, \tau_a}{d \, In \, \lambda} \tag{1.8}$$

where τ_a is the aerosol optical depth, α is the Angstrom parameter and λ is the wavelength. When α is equal or greater than 2, a fine mode aerosol is dominant. When α is near zero, the coarse mode aerosol is dominant.

AOD can be measured using either ground (sun photometer) or remotely sensed techniques. AERONET is known for harnessing ground measurements. It gives quality data on all aerosol column properties. However, it has a major limitation of few sites in developing and under-developed regions. The principle of remotely sensed technique is based on the ability of satellite to capture particulates in the atmosphere through the reflection and absorption of visible and infrared light. Remote sensing technique is available on some sites. For example, Aura/OMI are used to obtain aerosol optical depth at ground pixel resolution of 0.25° latitude/longitude grid and 1° latitude/longitude grid resolution; Meteor-3, TOMS and NIMBUS 7 are used to obtain aerosol optical depth at ground pixel resolution of 1° × 1.25° latitude/longitude grid resolution. Other satellite sites for obtaining AOD are Moderate Resolution Imaging Spectroradiometer (MODIS), Advanced Very High Resolution Radiometer (AVHRR), MEdium Resolution Imaging Spectrometer (MERIS), Polarization and Directionality of the Earth's Reflectances (POLDER) over ocean and Multi-angle Imaging SpectroRadiometer (MISR), Advanced Along Track Scanning Radiometer (AATSR), Total Ozone Mapping Spectrometer (TOMS), Ozone Monitoring Instrument (OMI), MODIS, Atmospheric Infrared Sounder (AIRS), TIROS Operational Vertical Sounder (TOVS) over land (NOAA 2015).

The main importance of the AOD is to determine the aerosol size distribution. In this study, the aerosol size distribution was obtained using the Multi-angle Imaging SpectroRadiometer (MISR). The MISR was launched in 1999 to measure the intensity of solar radiation reflected by the planetary surface and atmosphere. It operates at various directions, that is, nine different angles (70.5°, 60°, 45.6°, 26.1°, 0°, 26.1°, 45.6°, 60°, 20.5°) and gathers data in four different spectral bands (blue, green, red, and near-infrared) of the solar spectrum. The blue band is at wavelength 443 nm, the green band is at wavelength 555 nm, the red band wavelength 670 nm and the infrared band is at wavelength 865 nm. MISR acquire images at two different levels of spatial resolution; local and global mode. It gathers data at the local mode at 275 m pixel size and 1.1 km at the global mode. Typically, the blue band is to analyse coastal and aerosol studies. Blue band is higher at regions of increasing vegetation. The scope of the blue band may include ice, snow, soil or water. The blue band can therefore be divided into continental model blue band, desert model blue band, urban model blue band, biomass burning model blue band. The green band is to analyse Bathymetric mapping and estimating peak vegetation. The red band analyses the variable vegetation slopes and the infrared band analyses the biomass content and shorelines.

References

Ahmada, O. A. (2011). Modeling the dispersion of atmospheric pollutants dispersion using two dimensional advection diffusion equation, masters project submitted to University of Dar es Salaam, pp. 1–88.

Akinyemi, M. L., Emetere, M. E., & Usikalu, M. R. (2016). Virtual assessment of air pollution dispersion from anthropogenic sudden explosion. *American Journal of Environmental Sciences, 12*(2), 94–101.

Bai, F., Yang, Z., Huai, W., & Zheng, C. (2016). A depth-averaged two dimensional shallow water model to simulate flow-rigid vegetation interactions. *Procedia Engineering, 154,* 482–489.

Benson, D. A., Wheatcraft, S. W., & Meerschaert, M. M. (2000). Application of a fractional advection-dispersion equation. *Water Resources Research, 36*(6), 1403–1412.

Bobba, A. G., Vijay, P. S., & Lars, B. (2000). Application of environmental models to different hydrological systems. *Ecological Modelling, 125*(1), 15–49.

Bhatt, D., & Mall R. K. (2015). Surface Water Resources, Climate Change and Simulation Modeling. *Aquatic Procedia 4,* 730–738.

Choo-in, S. (2001). *Mathematical model for determining carbon monoxide and nitrogen oxide concentration in street tunnel.* M.Sc. Research, Thammasat University, Thailand. pp. 1–67.

Church, J. A., Gregory, J. M., Huybrechts, P., Kuhn, M., Lambeck, K, Nhuan, M.T., Qin, D., & Woodworth, P. L. (2001). Changes in sea level. In J. T. Houghton, Y. Ding, D. J. Griggs, M. Noguer, P. J. van der Linden & D. Xiaosu (Eds.), *Climate change 2001. The scientific basis* (pp. 639–693). Cambridge: Cambridge University Press.

Cullen, A. C., & Frey, H. C. (1999). *Probabilistic techniques in exposure assessment: A handbook for dealing with variability and uncertainty in models and inputs.* New York: Plenum.

Daitche, A., & Tamás, T. (2014). Memory effects in chaotic advection of inertial particles. *New Journal of Physics, 16*(073008), 1–35.

DBW. (2018). *Environmental modelling.* https://www.designingbuildings.co.uk/wiki/Environmental_modelling. Accessed February 24th, 2018.

Delft3D. (2018). *Flexible mesh—Environmental modelling.* https://www.deltares.nl/academy/delft3d-block-2a/. Accessed February 24th, 2018.

Devine, J. D., Sigurdsson, H., Davis, A. N., & Self, S. (1984). Estimates of sulfur and chlorine yield to the Atmosphere from volcanic eruptions and potential climatic effects. *Journal Geophysical Research, 89,* 6309–6325. https://doi.org/10.1029/JB089iB07p06309.

Edwards, J. B., McKinnon, A. C., & Cullinane, S. L. (2009). Carbon auditing the 'Last Mile': Modelling the environmental impacts of conventional and online non-food shopping. http://www.greenlogistics.org/SiteResources/ee164c78-74d3-412f-bc2a-024ae2f7fc7e_FINAL%20REPORT%20Online-Conventional%20Comparison%20%28Last%20Mile%29.pdf.

Emetere, M. E. (2014). Forecasting hydrological disaster using environmental thermographic modeling. *Advances in Meteorology, 2014,* 783718.

Emetere, M. E. (2016). Statistical examination of the aerosols loading over mubi-Nigeria: The satellite observation analysis. *Geographica Panonica, 20*(1), 42–50.

Emetere, M. E. (2017a). Investigations on aerosols transport over micro- and macro-scale settings of West Africa. *Environmental Engineering Research, 22*(1), 75–86.

Emetere, M. E. (2017b). Lightning as a source of electricity: Atmospheric modeling of electromagnetic fields. *International Journal of Technology, 8,* 508–518.

Emetere, M. E. (2017c). Impacts of recirculation event on aerosol dispersion and rainfall patterns in parts of Nigeria. *Global Nest Journal, 19*(2), 344–352.

Emetere, M. E. (2017d). Monitoring the 3-year thermal signatures of the Calbuco pre-volcano eruption event. *Arabian Journal of Geoscience, 10,* 94. https://doi.org/10.1007/s12517-017-2861-z.

Emetere, M. E., & Akinyemi, M. L. (2017). Documentation of atmospheric constants over Niamey, Niger: A theoretical aid for measuring instruments. *Meteorological Applications, 24*(2), 260–267.

Emetere, M. E. & Akinyemi, M. L. (2018). Sea level change in seven stations on the upper Atlantic: Implication on environments. *Journal of Physics: Conference Series.*

Emetere, M. E., Akinyemi, M. L., & Edeghe, E. B. (2016). A simple technique for sustaining solar energy production in active convective coastal regions. *International Journal of Photoenergy, 2016*(3567502), 1–11. https://doi.org/10.1155/2016/3567502.

EVO. (2018). *Environmental models*. http://www.evo-uk.org/at-the-outset/evo-cloud-services-portals/environmental-models/. Accessed February 24th, 2018.

FA. (2018). *Models—Aviation environmental tools suite*. https://www.faa.gov/about/office_org/headquarters_offices/apl/research/models/. Accessed February 24th, 2018.

Faccani, C., Rabier, F., Fourrie, N., Agustı́-Panareda, A., Karbou, F., Moll, P., et al. (2009). The impact of the AMMA radiosonde data on the French global assimilation and forecast system. *Weather and Forecasting, 24*, 1268–1286.

FES. (2018). *Environmental modelling*. https://www.fzp.czu.cz/en/r-9408-study/r-9495-study-programmes/r-9745-master-s-degree-programmes/r-9753-environmental-modelling. Accessed February 24th, 2018.

Ghumman, A. R., Yousry, M., Ghazaw, A. R., & Sohail, K. W. (2011). Runoff forecasting by artificial neural network and conventional model. *Alexandria Engineering Journal, 50*(4), 345–350.

Giuseppina, G. (2013). How far chemistry and toxicology are computational sciences? In *Methods and experimental techniques in computer engineering* (pp. 15–33). https://doi.org/10.1007/978-3-319-00272-9_2.

Gupta, V. R., & Jangid, R. A. (2011). The effect of bulk density on emission behaviour of soil at microwave frequencies. *International Journal of Microwave Science and Technology, 160129*, 1–6.

Guwahati IIT. (2014). *Advection-dispersion equation for solute transport in porous media*. https://nptel.ac.in/courses/105103026/32. Accessed August 20th, 2018.

Hämäläinen, R. P. (2015). Behavioral issues in environmental modelling—The missing perspective. *Environmental Modelling and Software, 73*, 244–253.

Hauduc, H., Neumann, M. B., Muschalla, D., Gamerith, V., Gillot, S., & Vanrolleghem, P. A. (2015). Efficiency criteria for environmental model quality assessment: A review and its application to wastewater treatment. *Environmental Modelling and Software, 68*, 196–204.

He, Q., Li, C., Geng, F., Yang, H., Li, P., Li, T., et al. (2012). Aerosol optical properties retrieved from Sun photometer measurements over Shanghai, China. *Journal of Geophysical Research, 117*(D16204), 1–8.

Holmes, N. S., & Morawska, L. (2006). A review of dispersion modeling and its application to the dispersion of particles: An overview of different dispersion models available. *Atmospheric Environment, 40*(30), 5902–5928.

Hughes, A., Jackson, C., Mansour, M., Bricker, S., Barkwith, A., Williams, A., et al. (2011, May). Integrated modelling within the Thames Basin: Examples of BGS work (Poster). In Cities, catchments and coasts: Applied geoscience for decision-making in London and the Thames Basin. London, UK. http://nora.nerc.ac.uk/14267/.

Jinduan, C., & Dominic, L. B, (2018). Forecasting hourly water demands with seasonal autoregressive models for real-time application. *Water Resources Research, 54*(2), 879–894.

Johnston, J. M., McGarvey, D. J., Barber, M. C., Laniak, G., Babendreier, J.E., Parmar, R., et al. (2011). An integrated modeling framework for performing environmental assessments: Application to ecosystem services in the Albemarlee Pamlico basins (NC and VA, USA). *Ecological Modelling, 222*(14), 2471–2484.

Lanzi, E. (2017). The economic consequence of outdoor air pollution. http://www.htap.org/meetings/2017/2017_May_2-3/presentations/10_TFIAM%20-%20Economic%20consequences%20of%20air%20pollution%20v2.pdf.

Liu, Y., Wang, Z., Wang, J., Ferrare, R., Newsom, R., & Welton, E. (2011). The effect of aerosol vertical profiles on satellite-estimated surface particle sulphate concentrations. *Remote Sensing of Environment, 115*(2), 508–513.

Logica. (2018). *Enhancing waterfall process through V-model software development methodology*. https://www.360logica.com/blog/enhancing-waterfall-process-through-v-model-software-development-methodology/. Accessed August 16th, 2018.

Min-Seop, A., & In-Sik, K. (2018). A practical approach to scale-adaptive deep convection in a GCM by controlling the cumulus base mass flux. *Climate and Atmospheric Science, 1,* 13.

Montibeller, G., & von Winterfeldt, D. (2015). Cognitive and motivational biases in decision and risk analysis. *Risk Analysis, 35*(7), 1230–1251.

National Research Council. (2007). *Models in environmental regulatory decision making.* Washington, DC: The National Academies Press. https://doi.org/10.17226/11972.

NOAA. (2015). http://www.esrl.noaa.gov/gmd/outreach/lesson_plans/. Accessed June 23rd, 2015.

OECD. (2014). *The cost of air pollution: Health impacts of road transport.* Paris: OECD Publishing. http://dx.doi.org/10.1787/9789264210448-en.

OECD. (2015). *The economic consequences of climate change.* Paris: OECD Publishing. http://dx.doi.org/10.1787/9789264235410-en.

Ogola, P. F. A. (2007). Environmental impact assessment general procedures. Paper pre-sented at short course II on Surface Exploration for Geothermal Resources. *Lake Naivasha: UNU-GTP and KENGEN*, Kenya.

Rotmans, J., & van Asselt, M. B. A. (2001). Uncertainty management in integrated assessment modeling: Towards a pluralistic approach. *Environmental Monitoring and Assessment, 69*(2), 101–130.

Samiksha, S. (2017). *Top 21 specialized branches of ecology—Discussed!* http://www.yourarticlelibrary.com/environment/top-21-specialized-branches-of-ecology-discussed/3801. Accessed December 30th, 2017.

Seddon, A. W. R., et al. (2013). Looking forward through the past: Identification of 50 priority research questions in palaeoecology. *Journal of Ecology, 102*(1), 256–267.

Shawn, M. L., Babendreier, J. E., & Thomas Purucker, S. (2009). Valuating uncertainty in integrated environmental models: A review of concepts and tools. *Water Resources Research, 45,* W06421. https://doi.org/10.1029/2008WR007301.

Shettle, E. P., & Fenn, R. W. (1979). Models for the aerosols of the lower atmosphere and the effects of humidity variations on their optical properties. Environmental research papers, AFGL-TR-79-0214, No. 676, pp. 1–23.

Sterman, J. D. (2002). All models are wrong: Reflections on becoming a systems scientist. *System Dynamics Review, 18*(4), 501–531.

Stockie, J. M. (2011). The mathematics of atmospheric dispersion modeling. *SIAM Review, 53,* 349–372.

Strong Todd, J., & Zundel Alan, K. (2014). Limitations of one-dimensional surface water models. *Journal of Undergraduate Research.* http://jur.byu.edu/?p=10582.

Sun T. Y., Gottschalk F., Hungerbuhler K., & Nowack B. (2014). Comprehensive probabilistic modelling of environmental emissions of engineered nanomaterials. *Environmental pollution, 185,* 69–76.

Thongmoon, M., McKibbin, R., & Tangmanee, S. (2007). Numerical solution of a 3-D advection-dispersion model for pollutant transport. *Thai Journal of Mathematics, 5*(1), 91–108.

WIKI. (2018). *Environmental niche modelling.* https://en.wikipedia.org/wiki/Environmental_niche_modelling. Accessed February 24th, 2018.

Wilby, R. L., Dawson, C. W., & Barrow, E. M. (2002). SDSM—A decision support tool for the assessment of regional climate change impacts. *Environmental Model and Software, 17,* 147–159.

Xiang, P., Geng, L., Zhou, K., & Cheng, X. (2017). Adverse effects and theoretical frameworks of air pollution: An environmental psychology perspective. *Advances in Psychological Science, 25*(4), 691–700.

Yoshioka, H., Koichi, U., & M, Fujihara. (2014). A finite element/volume method model of the depth-averaged horizontally 2D shallow water equations. *International Journal for Numerical Methods in Fluids, 75*(1), 23–41.

Zhang, S., Di, X., Li, Y., & Bai, M. (2013). One-dimensional coupled model of surface water flow and solute transport for basin fertigation. *Journal of Irrigation and Drainage Engineering, 139*(3), 1–8. https://doi.org/10.1061/(ASCE)IR.1943-4774.0000376.

Zhang, T., Ning, Xu, L., Guo, Y. H., & Yong, B. (2014). A global atmospheric contaminant transport model based on 3D advection-diffusion equation. *Journal of Clean Energy Technologies, 2*(1), 43–47.

Chapter 2
Introduction to Computational Techniques

Computational techniques are fast, easier, reliable and efficient way or method for solving mathematical, scientific, engineering, geometrical, geographical and statistical problems via the aid of computers. Hence, the processes of resolving problems in computational technique are most time step-wise. The step-wise procedure may entail the use of iterative, looping, stereotyped or modified processes which are incomparably less stressful than solving problems-manually. Sometimes, computational techniques may also focus on resolving computation challenges or issues through the use of algorithm, codes or command-line. Computational technique may contain several parameters or variables that characterize the system or model being studied. The inter-dependency of the variables is tested with the system in form of simulation or animation to observe how the changes in one or more parameters affect the outcomes. The results of the simulations, animation or arrays of numbers are used to make predictions about what will happen in the real system that is being studied in response to changing conditions.

Due to the adoption of computers into everyday task, computational techniques are redefined in various disciplines to accommodate specific challenges and how they can be resolved. Fortunately, computational technique encourages multi-tasking and interdisciplinary research. Since computational technique is used to study a wide range of complex systems, its importance in environmental disciplines is to aid the interpretation of field measurements with the main focus of protecting life, property, and crops. Also, power-generating companies that rely on solar, wind or hydro sources make use of computational techniques to optimize energy production when extreme climate shifts are expected. In this case, engineers, scientists and environmentalist are combining computational and meteorological dataset to address the challenge of understanding, characterizing, and predicting complex environmental systems. The most difficult task in computational techniques is understanding the computer programming language. A programming language is a formal language that highlights sets of instructions for execution. programming language is grouped by types namely Array languages, Assembly languages, Authoring languages, Constraint programming languages, Command line interface languages, Compiled lan-

© Springer Nature Switzerland AG 2019
M. E. Emetere, *Environmental Modeling Using Satellite Imaging and Dataset Re-processing*, Studies in Big Data 54,
https://doi.org/10.1007/978-3-030-13405-1_2

guages, Concurrent languages, Curly-bracket languages, Dataflow languages, Data-oriented languages, Decision table languages, Declarative languages, and Embeddable languages. Array language is programming language that is used to convert operations from scalars to vectors, matrices, and higher-dimensional arrays. A typical example of modern languages that supports array language includes the following: Fortran 90, Mata, MATLAB, Analytica, TK Solver (as lists), Octave, R, Cilk Plus, Julia, Perl Data Language (PDL) and the NumPy extension to Python. An assembly language is a group of low-level programming languages used by microprocessors and other programmable devices to implement symbolic representation of machine code needed to program a given CPU architecture. The type of assembly language includes: Complex Instruction-Set Computer (CISC), Reduced Instruction-Set Computer (RISC), Digital Signal Processor (DSP) and Very Long Instruction Word (VLIW).

Authoring language is a notation used to control the appearance and functionality of webpages when displayed in a browser. Example of authoring language are DocBook, DITA, PILOT, TUTOR, Bigwig, Chamilo, Hollywood (Hollywood Designer graphical interface), Learning management system, SCORM, Web design program, XML editor and Game engine. Constraint programming is a type of programming paradigm that displays variable in the form of constraints. Libraries that accepts constraint programming are Artelys Kalis, C++, Java, Python library, FICO Xpress module, Cassowary, Smalltalk, Ruby library (LGPL), CHIP V5, Choco, Java library, Comet, Cream, Java library (LGPL), Disolver, Facile, OCaml library (CC0 1.0), finite-domain, Haskell library (MIT), Gecode, Google OR-Tools, JaCoP, Java library, LINDO MonadicCP, Haskell library (BSD-3-Clause) etc. Command line interface languages are languages that are used to interact with a computer program where the user issues commands to the program in the form of successive lines of text. The successive lines of text are called command lines. The command lines are executed in the shell of operating system. The shell of operating system includes AmigaOS (Amiga CLI/Amiga Shell), Unix OS (Bourne shell, Almquist shell, Debian Almquist shell).

Bash, Korn shell, Z shell, C shell, TENEX C, Emacs shell, rc shell rc, Standalone shell and Remote shell), Microsoft Windows (CMD.EXE, Windows PowerShell, Hamilton C shell, 4NT, Recovery Console), DOS(COMMAND.COM, 4DOS, NDOS and GW-BASIC), OS/2 (CMD.EXE, Hamilton C, and 4OS2), IBM OS/400 (AS/400 Control Language, iSeries QSHELL) Apple (Apple DOS/Apple ProDOS) and Mobile devices (DROS, Java ME platform).

Compiled languages is a programming language whose implementations are typically compilers (translators that generate machine code from source code), and not interpreters (step-by-step executors of source code). Examples of compiled programming language are ALGOL, BASIC, C, D, CLEO, COBOL, Cobra, Crystal, Eiffel, Fortran, Go, Haskell, Haxe, JOVIAL, Julia, LabVIEW, G, Pascal, SPITBOL, Visual Foxpro and Visual Prolog. Embeddable languages are programming language that supports scripting in real-time systems. Example of embedded language includes Python, C++ and Java.

The utmost goal of researchers or modeler is to develop an experimentally driven computer model that generates accurate predictions of environmental scenarios e.g. Numerical Weather Prediction (NWP)model. A typical example of an experimentally driven computer model is the Numerical Weather Prediction (NWP) developed by the National Oceanic and Atmospheric Administration (NOAA). NWP is a form of day-to-day weather model data. NWP focuses on taking current observations of weather and processing these data with numerical computer models to forecast the future state of weather. In this case, current weather or meteorological observations serve as input to the numerical computer models through a process known as data assimilation to produce outputs of temperature, precipitation, and hundreds of other meteorological elements from the oceans to the top of the atmosphere (NOAA 2018). The NWP data are Global Ensemble Forecast System (GEFS), Global Forecast System (GFS), Climate Forecast System (CFS), North American Multi-Model Ensemble (NMME), North American Mesoscale (NAM), Rapid Refresh (RAP) and Navy Operational Global Atmospheric Prediction System (NOGAPS). The assimilation data of the NWP is obtained from the Global Data Assimilation System (GDAS). Before the usage of assimilation data in environmental model, optimum interpolation (OI) method was used. The OI was introduced by L. S. Gandin (István et al. 2013). Its use was discontinued because of the limitation/drawbacks. The variational method replaced the OI until it was also discontinued mainly because of some programing and computational difficulties it encountered.

In the early days of understanding the role of computers in fostering research, some profession e.g. management, economics, biology etc. makes use of statistical packages only. In recent time, advancement of computers and computer application has brought about more sophisticated computer packages for solving problems (Emetere and Sanni 2015). In this chapter, we shall discuss on general outline on computational techniques, open-software packages and libraries. In science and engineering, computational technique is beyond using computer application. It entails restructuring related mathematical or physical theories to solve or optimize a specific process.

2.1 General Outline of Computational Techniques

What generally comes to mind when computational technique is mentioned is computer software. Computer software are structured algorithms or codes that implement a specific task. Hence, the three broad classification of computer software namely system software, application software and programming software. System software are made-up of operating system (Microsoft Windows, Mac OS X, and Linux) device drivers (Bios, motherboard drivers, hardware divers, virtual device drivers, sound cards etc.), servers and software components. Application software (AS) is used for attaining specific tasks. The types of AS available in the market includes licensed, sold, freeware, shareware, and open source software. In this chapter, the emphasis shall be on the open source (since most environmental models are open source).

Fig. 2.1 Computational technique for data mining (Chandra and Chinmayee 2012)

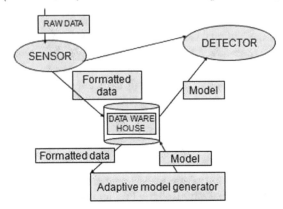

Fig. 2.2 Computational technique for materials (Korolkovas 2016)

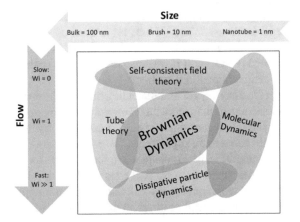

Programming software are mini-codes or macros used for writing programs through tools such as editors, linkers, debuggers, compilers/interpreters etc.

Computational techniques may be seen as a very broad concept in modern research. The early type of computational technique is termed computer aided algebraic systems. In this technique, algebraic related problems are solved using algorithms, codes, charts and special syntaxes. However, as research expanded, there were need for more application software for special task. In environmental modelling, there are general and specific application software. The general application software performs tasks as speculated by the modeler discretion. Example of general application software in environmental modelling includes MATLAB, MATHCAD, Origin, Homer, GNU plot, FreeMat, Octave, Microsoft Excel, Magma, Maple, R package, Mathematical, SageMath, PolyMath, SMath, COMSOL etc. The various computational techniques are shown pictorially in Figs. 2.1 and 2.2.

Fig. 2.3 Pictorial illustration of computational technique

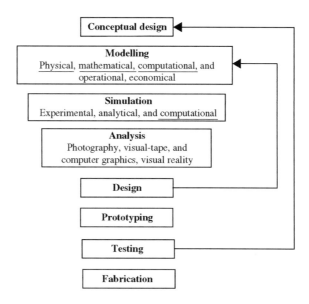

The specific application software (Captera 2017) for environmental modelling includes SimaPro or GaBi (for assessing Carbon Footprints), Qual2K (river and stream water quality and process interaction model), Hec-HMS (for hydrologic modeling system), WaterCAD (for designing and planning hydrologic systems), ADORA or AERMOD (for monitoring air pollution), EMEX (for managing incidents and track corrective and preventative actions), Enviro (for gathering, managing, and displaying lab and field data for water, soil, and air), SmartData (for investigating environmentally contaminated sites), Accuvio (for monitoring carbon print and greenhouse gases) etc. In recent time, it has been proven that some environmental application software has their flaws (Patwardhan 2016). This development is quite worrisome because beginners and young professionals in environmental studies may be building on faulty foundation.

Based on the aforementioned, environmental scientists are advised to use either the general application software or adopt open-source application/library to prevent uncontrollable error-prone analysis/research. The outline of the computational technique will be explained using the pictorial illustration in Figs. 2.1, 2.2 and 2.3.

The fabrication stage is referred as the developmental procedures where the modeler or researcher decides on what kind of computational technique would be appropriate for the model. Most modeler at this stage decides to adopt mathematical methods. Mathematical method is a tool for solving abstract and real problems. Mathematical method is fondly used in science, engineering, social science and humanities. The list of method that are usually considered in mathematical methods includes: Infinite series, power series, Complex numbers, Integral transform, Wavelet transform, Fourier transform spectroscopy, Harmonic analysis, Linear algebra, Partial differentiation, Multiple integrals, Vector analysis, Fourier series and transforms, Ordinary differential equations, Calculus of variations, Linear function, Quadratic function, Cubic function, Quartic function, Cabibbo-Kobayashi-Maskawa matrix, Density matrix, Fundamental matrix, Fuzzy associative matrix, Gamma matrices, Gell-Mann, Hamiltonian matrix, Wall polynomial, Wangerein functions, Weber function, Weierstrass function, Weisner's method, Whittaker function, Wilson polynomial, Irregular matrix, Overlap matrix, S-matrix, State transition matrix, Substitution matrix, Z-matrix, Quintic function, Sextic function, Tensor analysis, Special functions, Schubert polynomial, Schur polynomial, Selberg integral, Sheffer polynomial, Slater's identities, Stieltjes polynomial, Stieltjes–Wigert polynomials, Strömgren function, Struve function, Legendre function, Bessel function, Hermite function, Laguerre function, Partial differential equations, Functions of a complex variable, Integral transforms, Gamma function, Barnes G-function, Beta function, Digamma function, Polygamma function, Incomplete beta function, Incomplete gamma function, K-function, Multivariate gamma function, Student's t-distribution, Probability and statistics, Two-sided Laplace transform, Mellin transform, Laplace transform, Fourier transform, Fourier series, Sine and cosine transforms, Hartley transform, Short-time Fourier transform, Celine's polynomial, Charlier polynomial, Pafnuty Chebyshev, Chebyshev polynomials, Painlevé function, Painlevé transcendents, Poisson–Charlier polynomial, Pollaczek polynomial, Cyclotomic polynomials, Rectangular mask short-time Fourier transform, Gegenbauer polynomials, Gottlieb polynomial, Gould polynomial, Gudermannian function, Chirplet transform, Fractional Fourier transform (FRFT), Hankel transform, Hall polynomial, Hall–Littlewood polynomial, Hankel function, Heine functions, Racah polynomial, Riccati–Bessel function, Riemann, zeta-function, Rodrigues formula, Rogers–Askey–Ismail polynomial, Rogers–Ramanujan identity, Rogers–Szegő polynomials, Hermite polynomials, Heun's equation, Horn hypergeometric series, Hurwitz zeta-function, Boubaker polynomial etc. However, some modeler would want to generate their own brand of mathematical methods.

Secondly, the modeler choose what kind of programming language would be appropriate to solve the mathematical method that has been adopted. The common programming language used in research in modern times include: Python, Q# (Microsoft programming language), C, C–, Java, C++, C#, MATLAB, Mathcad, Visual Fortran.

Visual FoxPro, JavaScript, JCL, Jython, MATH-MATIC, Visual J++, Visual J#, Mathematica etc. Few researcher or modeler used one or more of the aforementioned programming language.

As shown in Fig. 2.3, it is mandatory for the researcher or modeler computationally validate the solved problem using dataset. If the modeler is not satisfied with the outcome, he checks his computational procedures once more until he/she can figure-out how to obtain an accurate outcome. Once the modeler is satisfied with the testing of the computational work, he moves on to the prototyping of the computational work. Most modeler stops at the testing stages because of funds or technical know-how. The prototype is sent to research centers or industries for feedback purposes.

Once the feedback comes out from the sources, the modeler goes on to the designing stage. At this point the computational work comes out as a licensed or open-source application. The modelling, analysis and simulation stages are performed by the end-users.

2.2 Open Source Scientific Packages

The open source scientific packages are generally referred to as computer software that are licensed either under the free software licenses or the open-source licenses. However, the term 'open-source' may not necessary be without financial involvement as there are commercial open-source applications that are linked to business models e.g. AdaControl, Sun Studio, Dolibarr, Openbravo, EyeOS, Kaltura, LucidWorks, Zenoss Core, OrangeHRM, Qt, Talend Open Profiler etc. The idea of the open source can be illustrated as shown in Fig. 2.4. The most important factor of the open source is the possibility of low transaction costs which may only occur if the supportive library of the open source application is not available on the local computer.

There are notable scientific open source software applications that have gained relevance in its application in environmental studies. Tabula is a scientific package that allows users to extract data from pdf into a CSV spreadsheet using a simple and easy-to-use interface. This package is particularly useful when there is large data in the pdf format. However, the shortcoming of this software is that the processing speed is slow. Dakota is another freely available engineering software framework for large-scale optimization and uncertainty analysis. Dakota software's advanced

Fig. 2.4 The ideals of open source (Wilbanks 2013)

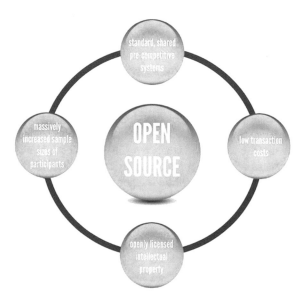

parametric analyses enable design exploration, model calibration, risk analysis, quantification of margins and uncertainty with computational models. This features are very important in environmental engineering because it helps to improve on the accuracy of predictions. The summarized list of scientific software includes, 3D Slicer, AMPHORA2, Ascalaph Designer, Bioconductor, BioModels Database, Biskit, Brian Simulator, ChemTool, Cn3D, DataMelt, EMBOSS, Emergent, GenMAPP, GENtle, GIMIAS, Gnaural, Gwyddion, HMMER, ImageJ, InVesalius 3, LabKey, MDynaMix, OpenMS, OsiriX Imaging Software, PathVisio, QuteMole, RasMol and OpenRasMole, RDQA, Spatiotemporal Epidemiological Modeler (STEM), UGENE, Virtual Cell (VCell), XDrawChem, ZygoteBody etc.

In recent times, researchers or modelers seek to build their own open source application. Ibrahim (2010) in his blog simplified the foundational requirements to begin open source project with six salient questions i.e.

i.   Can we financially sponsor the project? Do we have an internal executive champion?
ii. Is it possible to join efforts with an existing open source project?
iii. Can we launch and maintain the project using the OSS model?
iv. What constitutes success? How do we measure it?
v.   Will the project be able to attract outside enterprise participation (from the start)?
vi. Is there enough external interest to form and grow a developer community?

The key to open-source development is the ability of the researcher to understand the licenses of the programs he/she will be using to create his/her open source application. Compatibility issues between licenses can originate when you are trying to include open source code as a library in your existing project. Hence, it is very important for a modeler to understand in clear terms the details of each licenses to avoid copyright infringement. For example, Seher (2017) explained that software package licensed under an Apache 2.0 license are compatible with software license of GNU General Public License, version 3 (GPLv3) because Apache 2.0 terms are covered by GPLv3 because both licenses have same patent usage protection. The GPLv3 license reads: the source code must be made public whenever a distribution of the software is made; modifications of the software must be released under the same license; changes made to the source code must be documented; If patented material was used in the creation of the software, it grants the right for users to use it. If the user sues anyone over the use of the patented material, they lose the right to use the software. However, GPLv2 license is not compatible with Apache 2.0 because of patent grant clause that is missing. License compatibility can therefore be considered as a subtle danger that must be considered carefully. Some authors have given insight on how to avoid copyright infringement. For example, Välimäki (2005) reported a schematic sketch of license compatibility as presented in Fig. 2.5.

The term derivative work in Fig. 2.5 refers to as everything that uses the source code in any way possible. For example, Berkeley Software Distribution (BSD) 2-clause supports derivative work. However, 3-clause of BSD do not support derivative work because it states *"the names of the author and contributors can't be used to promote products derived from the software without permission"*. This gives rise

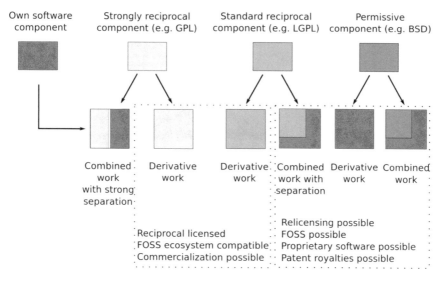

Fig. 2.5   License compatibility issues (Välimäki 2005)

to the question: "what is open source license". Readers should take a little time to go through the definition of open source by Open Source Initiative (OSI 2018). The four component of the open source licenses are: software can be modified, used commercially, and distributed; software can be modified and used in private; a license and copyright notice must be included in the software; software authors provide no warranty with the software and are not liable for anything.

The second key to open-source development is the organization of the new project. Flory (2018) reported in his article that managing an open source project is a challenging work, and the challenges grow as the project grows because at every developmental stages of the project, it is expected that the project meet different requirements and span multiple repositories. The author gave three tips for organizing an open source project. The three tips are: bring development discussion to issues and pull requests (sincerity and transparency is advised); Set up kanban-style project boards (projects boards are repository boards that are used in a single repository) and Organization (boards for use in a GitHub organization across multiple repositories); Build project boards into your workflow. The workflow is pictorially defined in the chart presented in Fig. 2.6. For in-depth reading, readers are advised to go to the link presented in the reference section.

The third key to open-source development is the strict adherence of author or modeler to maintain openness. This process helps expert to modify features in the project. Sometimes, developers express themselves quite bluntly, so the modeler must be emotionally balance to take the message and discard the insults. Also, inexperience volunteers may cause set-backs in the open source development. In this case, suggestions or contributions from experts are out rightly not in line of discuss.

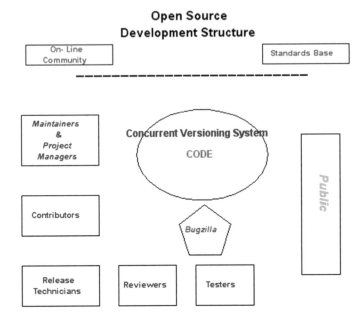

Fig. 2.6 Open source development structure (Tom 2004)

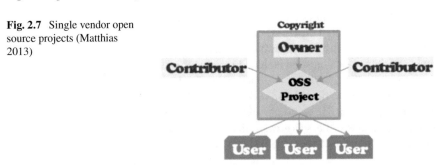

Fig. 2.7 Single vendor open source projects (Matthias 2013)

The ability of the modeler to decipher between useful and wasteful contributions is very important for the onward progress of the open source project.

Matthias (2013) stated clearly that open source software is not only about programming code but it is a carefully organized process that is hinged on a systematic order. In the light of his explanation, Matthias (2013) identified four main organizations within the open source community namely: single vendor open source projects; development communities; user communities; and open source competence centers. The pictorial chart that explicitly describes the mode of operations is presented in Figs. 2.7, 2.8, 2.9 and 2.10.

In environmental research, there are about a thousand open-source scientific packages that are used currently in the field. NASA (2018) listed several open-source that can be operated by novice and professionals. Most professional open-source appli-

Fig. 2.8 Development communities (Matthias 2013)

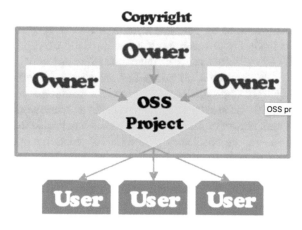

Fig. 2.9 User communities (Matthias 2013)

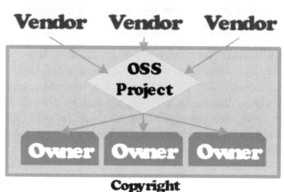

Fig. 2.10 Open source competence centers (Matthias 2013)

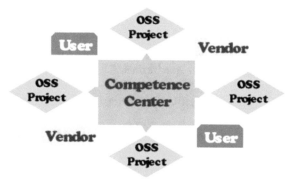

cation packages are written in Python or C++ programming language. One of the many reasons for adopting Python or C++ in open-source application is because of its flexibility to accommodate the 'big data' concept. Data science space adopts the C++ language because of the nature of its operation.

Open-source applications that are written in Python, Java or C++ language are adjudged appropriate: when complex machine learning algorithms are involved; when the dataset is in terabyte or petabyte; when working on deep learning and deep neural networks. However, researchers have noted that few open-source application packages are obsolete. Upasani (2016) noted that open-source application/libraries need to embrace digital technologies and library management systems (LMS) in order to work smart and achieve more with less. Higgs (2016) believes that open source software is unsupported, unsustainable and unreliable. Hence, big companies do not patronize open-source software. ConnectUS (2018) also highlighted the disadvantages of open-source software namely vulnerability to malicious users, not user-friendly as commercial versions and do not come with extensive support.

The preferred open-source application used by the author is the CERN-Root software. ROOT is an object-oriented program and library developed by European Organization for Nuclear Research (CERN). It was originally designed for particle physics data analysis but it is also used in other applications such as astronomy and data mining. Root are used for plotting histograms and graphs, curve fitting, statistical analysis, data analysis, matrix algebra, four-vector computations, multivariate data analysis, image manipulation, 3D visualizations (geometry), creating files in various graphics formats, interfacing Python and Ruby code in both directions and interfacing Monte Carlo event generators.

2.3 Open-Source Library

Library is a collection of non-volatile resources used by computer programs, often to develop software. Libraries are used for software development to enhance the software to perform specific task. The major computer libraries are written in terms of language e.g. Multi-language, C, C++, Delphi, .NET Framework languages (C#, F#, VB.NET and PowerShell), Fortran, Java, Scala, Perl, Python, Groovy, XNUMBERS, INTLAB etc.

Open-source library has almost the same shortcoming as open-source application. Since most open-source allows anyone to interact with its source codes, the open-source library can be modified to suite any task. For example, Igor (2017) modified added commits, contributors count and other metrics from Github to enhance the proxy metrics for Python library popularity.

Open-Source library are designed to perform certain function. Also, library can be built upon one another. For example, NumPy is a Python library that provides a fundamental framework where scientific computation stack is built. The main functionality of SciPy library is built upon NumPy. Most open-source libraries are low-level tool i.e. it requires more codes by the modeler to advance its status to a high-level tool. For example, the matplotlib library cannot exist alone, except it is enhanced by the stack of NumPy, SciPy and Pandas Python library.

One of the preferred open-source library-used by the author is the OpenCV. OpenCV is an open and free source computer vision library that is released under a BSD license. It has C++, C, Python and Java interfaces and supports Windows, Linux, Mac OS, iOS and Android. OpenCV is used for real-time applications like image processing, matrix algebra, four-vector computations, multivariate data analysis, image manipulation, 3D visualizations (geometry), creating files in various graphics formats, interactive art, mines inspection evaluation, stitching maps on the web or through advanced robotics. OpenCV is enabled with OpenCL to boost its hardware acceleration of the underlying heterogeneous computational platform.

The C++ libraries includes: Boost, GSL, BDE, Dlib, JUCE, Loki, Reason, yomm2, Folly, Abseil, cxxomfort, libsourcey, OnPosix, Ultimate++, CAF, cpp-mmf, CommonPP, Better Enums, Smart Enum, nytl, SaferCPlusPlus, fcppt, bit-field.h, composite_op.h, Yato, Kangaru, yaal, rpnx-serial,libnavajo, C++ RESTful framework, C++ REST SDK, cpr, cpp-netlib, cpp-redis, tacopie, Boost.Asio, Boost.Beast, gsoap,POCO, omniORB, ACE, TAO, wvstreams, Unicomm, restful_mapper, zeromq, curlpp, Apache Thrift, libashttp, Simple C++ REST library, libtins, PcapPlusPlus, HTTPP, The Silicon C++14 Web Framework, ngrest, restc-cpp, OpenDDS, Breep, uvw, rest_rpc, EasyHttp, nghttp2,Dear ImGui, FLTK,nana[, WxWidgets, OWLNext,tiny file dialogs, Switch, glibmm, gtkmm, goocanvasmm, libglademm, libgnomecanvasmm, webkitgtk, flowcanvas, evince, Qt, qwtplot3d, qwt5, libdbusmenu-qt, QuickQanava, QuickProperties, SFML (Simple and Fast Multimedia Library),SDL (Simple DirectMedia Layer), SIGIL (Sound, Input, and Graphics Integration Library), Cinder, openFrameworks, cairomm, nux, pangomm, gegl, stb, Adobe/boost GIL, GraphicsMagick, Skia Graphics Engine, enwiki:Skia_Graphics_Engine, Anti-Grain Evolution, plotutils, libraw, openexr, qimageblitz, imagemagick, djvulibre, poppler, SVG++, id3lib, taglib, opencv, dlib, ITK, OTB, Vulkan, OpenGL, bgfx, Ogre3D, Diligent Engine, GLEW, GLAD, Epoxy, GLFW, GLM, hlsl++, assimp, VTK, Magnum, Irrlicht, Horde3D, Visionaray, Open CASCADE, OpenSceneGraph, EntityX, Anax, EntityPlus, EnTT, BOX2D, stats++, StatsLib, alglib, ArrayFire High Performance Computation Library, GNU MP bignum C++ interface, BigNumber, Boost.Multiprecision, Boost.Math.Special Functions and Statisticalistributions, Boost.Random, NTL - A Library for doing Number Theory, cpp-measures, G + Smo cross-platform library for isogeometric analysis, Exact floating-point arithmetic library, Boost.uBLAS, Eigen, Armadillo, Blitz++, IT++, Dlib - linear algebra tools, Blaze, ETL, DecompLib, OptimLib, Boost.Graph, LEMON, OGDF—Open Graph Drawing Framework, NGraph—a simple (Network) Graph library in C++, GTpo, cln, Dlib—machine learning tools, MLPACK—machine learning package, Shogun—large scale machine learning toolbox, CGAL—Computational geometry algorithms library, Wykobi—Computational geometry library, PCL—Point Cloud library, yasmine—C++11 UML state machine framework, libxml++, pugixml, tinyxml, tinyxml2, Xerces, gSOAP, ai-xml, json, ArduinoJson, jsonme–, ThorsSerializer (JSON/YAML Input Output Streams), Json-Box, jsoncpp, zoolib, JOST, CAJUN, libjson, nosjob, rapidjson, jsoncons, JSON++, qjson, json-cpp, jansson, json11, JSON Voorhees, jeayeson, ujson, minijson, jios (JSON Input Output Streams), Botan, gnutls, openssl, crypto++, TomCrypt etc.

The Python libraries includes: ADOdb, AppJar, Beautiful Soup (HTML parser), CGAL, CheetahTemplate, Construct (python library), Cubes (OLAP server), Genshi (templating language), Gensim, IronPython, Jinja (tmplate engine), Kamaelia, Kid (templating language), Kivy (framework), Natural Language Toolkit, Pickle (Python), PLWM, PyEphem, Pygame, Pyglet, PyGObject, PyGTK, PyObjC, PyQt, PySide, Python Imaging Library, Python Robotics, Python-Ogre, RDFLib, Redland RDF Application Framework, Requests (software), RPyC, SimpleITK, SimPy, Sound Object (SndObj) Library, Soya3D, SpaCy, SQLAlchemy, SQLObject, Storm (software), Tkinter, Topsite Templating System, Twisted (software), VPython, WxPython, XDMF, Pipenv, PyTorch, Caffe2, Pendulum, Dash, PyFlux, Fire, imbalanced-learn, FlashText, Luminoth, Scrapy, Pillow, NumPy, SciPy, matplotlib, Scapy, pywin32, nltk, nose, SymPy, IPython etc.

Open source library uses some open source software like Linux, Apache Web Server, OpenOffice, GIMP, Audacity, and Firefox browsers. Koha and Evergreen are referred to as integrated library systems (ILS). They are the most popular libraries and both are licensed under a GNU General Public license. Opens source application depends on open source libraries to perform certain operation or features. Example of the dependent open source applications are as follows: OpenCog, OpenCV, TREX, ROS, YARP, FreeCAD, BIM, LibreCAD, Blender, flightgear, SimPy, Scribus, Bitcoin Core, Bonita Open Solution, CiviCRM, Compiere, Cyclos, Dolibarr, ERPNext, GnuCash, HomeBank, iDempiere, Ino erp, jFin, JFire, KMyMoney, LedgerSMB, metasfresh, Mifos, Odoo, Openbravo, OrangeHRM, Postbooks, QuickFIX, Quick-FIX/J, SQL Ledger, SugarCRM, Tryton, TurboCASH, Wave Accounting, ZipBooks, Evergreen, Koha, NewGenLib, OpenBiblio, PMB, refbase, Darktable, digiKam, GIMP, Inkscape, Krita, LightZone, RawTherapee, Chemistry Development Kit, JOELib, OpenBabel, P-GRADE Portal, CellProfiler, Endrov, FIJI (software), Ilastik, ImageJ, IMOD, ITK, KNIME, VTK, 3DSlicer, Abalone, Ascalaph Designer, GRO-MACS, LAMMPS, MDynaMix, NAMD, NWChem, Avogadro, BALLView, Jmol, Molekel, MeshLab, PyMOL, QuteMol, RasMol, Ninithi, CP2 K, LimeSurvey, CMU Sphinx, Emacspeak, ESpeak, Festival Speech Synthesis System, Modular Audio Recognition Framework, NonVisual Desktop Access, Text2Speech, Dasher, Gnopernicus Virtual Magnifying GlassEnvironment, Konstanz Information Miner (KNIME) OpenNN, Orange, RapidMiner, Scriptella ETL, Weka, JasperSoft, ParaView, VTK, ResourceSpace, ApexKB, Lucene, Nutch, Solr, Xapian, Elasticsearch, Konstanz Information Miner (KNIME), Pentaho, SpagoBI, Talend, OpenAFS, Tahoe-LAFS, CephFS, OpenX, Asterisk. Ekiga, FreePBX, FreeSWITCH, Jitsi, QuteCom, Enterprise Communications System sipXecs, Twinkle, Ring, Tox, Geary, Mozilla Thunderbird, GNU Queue, HTCondor OpenLava, pexec, Apache Axis2, Apache Geronimo, Bonita Open Solution, GlassFish, Jakarta Tomcat, JBoss Application Server, ObjectWeb JOnAS, TAO, Enduro/X, Akregator, Liferea, RSS Bandit, RSSOwl, Sage, Popcorn Time, qBittorrent, Drupal, Liferay, Oxwall, Sun Java System Portal Server, uPortal, FreeNX, OpenVPN, rdesktop, Synergy, VNC, Remmina, Brave, Chromium, Firefox, Midori, Tor Browser, Waterfox, SeaMonkey, Cheese, Guvcview, cURL, HTTrack, Wget, Apache Cocoon Apache, AWStats, BookmarkSync, Chero-kee, curl-loader, FileZilla, Hiawatha, HTTP File Server, lighttpd, Lucee, Nginx, NetKernel, Qcodo, Squid, Vaadin, Varnish, XAMPP, Zope, ATutor, Chamilo, Claro-line, DoceboLMS, eFront, FlightPath, GCompris, Gnaural, H5P, IUP Portfolio, ILIAS, Moodle, OLAT, Omeka, openSIS, Sakai Project, SWAD, Tux Paint, Uber-Student, KGeography, Kiten KVerbos WINE, CyberBrau, Pencil2D, Pivot Animator, Synfig, Tupi (formerly KTooN), OpenToonz, Blender, OpenFX, Seamless3d, Pen-cil2D, SWFTools, Eye of GNOME, F-spot, Geeqie, Gthumb, Gwenview, Kphotoal-bum, Opticks, Dr. DivX, FFmpeg, MEncoder, OggConvert, Avidemux, AviSynth, Blender, Cinelerra, DScaler, DVD Flick, Flowblade, Kaltura, Kdenlive, Kino, LiVES, Natron, OpenShot Video Editor, Pitivi, Shotcut, VirtualDub, VirtualDub-Mod, VideoLAN Movie Creator, Avidemux, HandBrake, FFmpeg, Apache OpenOf-fice, Calligra Suite, LibreOffice, Chandler, KAddressBook, Kontact, KOrganizer, Mozilla Calendar, Novell Evolution, OpenSync, Project.net, TeamLab, Bugzilla, Mantis, Mindquarry, Redmine, Trac, Bison, CodeSynthesis XSD, CodeSynthe-sis XSD/e, Flex lexical analyser, Kodos, Open Scene Graph, OpenSCDP, php-CodeGenie, SableCC, SWIG xmlbeansxx, YAKINDU Statechart Tools, Doxygen, Mkd, Natural Docs, Autoconf, Automake, BuildAMation, CMake, GNU Debug-ger, Memtest86, Xnee, BOINC, Electric Sheep, XScreenSaver, MyDLP, dvdisas-ter, Foremost, PhotoRec, TestDisk, Mydiamo, Coyote Linux, Firestarter, IPFilter, ipfw, iptables, M0n0wall, PeerGuardian, PF, pfSense, Rope, Shorewall, SmoothWall, Untangle, Vyatta, Java Astrodynamics Toolkit (GPL), General Mission Analysis Tool (NASA Open Source Agreement), OREKIT (ORbits Extrapolation KIT) (Apache License), Satellite tracking and orbit prediction (GPL), Orbit Reconstruction Simula-tion and Analysis (GPL), Asteroid Orbit Determination and Propagation (GPL), Lib-nova (LGPL), Open-Source, Extensible Spacecraft Simulation And Modeling Envi-ronment (GPL), Distributed Spacecraft Attitude Control System Simulator (GPL), Solar Sail structure and flight simulator (GPL), SaVi satellite constellation visualizer (BSD License), Rocket Workbench Project (GPL), CpropepShell. Compute propel-lant performance, BRL-CAD, Blender, Blender CAD, Procad, OpenSCAD, Python CAD, VARKON, OpenCASCADE, FreeCAD, Archimedes, Wikipedia Free CAD Software Listing, Linux.org CAD/CAM Software Listing, NASA Vision Workbench (NOSA license), wikiCalc (GPL), Dia (GPL), DUNE (GPL with runtime exception), Impact (GPL), Code_Aster, SALOME (LGPL), Elmer, Gmsh, OpenFVM, Palabos, Calculix, Package of Additional Octave Libraries, ASCEND modelling environ-ment, OpenDX (IBM), Freshmeat.net Visualization Software Listing, VisIt (BSD), EngLab (GPL), SciLab (CeCILL license), WorldWind (NOSA), Numpy/Scipy, OpenCog, AForge.NET, TREX, ROS, YARP, Blender, flightgear, Chemistry Devel-opment Kit, JOELib, CellProfiler, Endrov, FIJI, Ilastik, ImageJ, IMOD ITK, KNIME, VTK, 3DSlicer, Abalone, Ascalaph Designer, GROMACS, LAMMPS, MDynaMix, NAMD, NWChem, Avogadro, BALLView, Jmol, Molekel, MeshLab, PyMOL, QuteMol, RasMol, Ninithi, CP2 K, CMU Sphinx, Emacspeak, ESpeak, Bullet, AwayPhysics, Bullet-ANE, ammo.js, Physijs, AmmoNext, Bullet.js, JBullet, ODE, Bounce, nphysics, Velocity Raptor, Box2D, Nape, GoblinPhysics, verlet-js, Physic-sJS, Matter.js, p2.js, Coffee Physics, JPE, APE, Chipmunk2D, glaze, ImpulseEngine,

JelloAS3, JelloHx, Jello-Physics, JelloSwift, JigLib, Moby, Newton-Dynamics, OimoPhysics, qu3e, Tokamak, DynaMo, ReactPhysics3D, Chrono Engine, Position Based Dynamics, SPlisHSPlasH etc.

Modeler sometime desire to create his/her own libraries. The step is quite easy if the preliminary steps are diligently executed. For example, there must be an interface to the proposed library. The header file to the proposed library should contain definitions for everything exported by your library. This includes: function prototypes with comments for users of your library functions; definitions for types and global variables exported by your library. The modeler is expected to write a "boiler plate" code that enables the preprocessor to include the 'proposedlib.h' file one time. Aside creating the interface, the modeler is expected to design the implementation flow chart of the library. This exercise is achieved by creating a proposedlib.c file that #includes "proposedlib.h". The next step after creating implementation code is creating a library object file that can be linked with programs that can access the proposed library code. Alternatively, modeler may wish to create a shared object file from many.o files that can be linked with programs that want to use the proposed library code. However, before you share file, be sure of the license compatibility issues. The file sharing can be achieved by linking the proposed.c file with the library object file. An example of the.c file linking in C language is presented below.

" gcc test.c mylib.o

OR to link in libmylib.so (or libmylib.a):

gcc test.c -lmylib

OR to link with a library not in the standard path:

gcc test.c -L/home/newhall/lib –lmylib"

An example of the.c file linking in C ++ language is presented below.

"INC = -I ./Headers

g++ main.o proposedlib.o

g++ $(INC) –c main.cpp

g++ $(INC) –c proposedlib.cpp

rm proposedlib.o main.o a.out

make

./executable"

An example of linking your created module to the main module in python language is presented below.

```
"import proposedlib
if __name__ == "__main__":
    import sys
    fib(int(sys.argv[1]))"
```

An example for linking the C++ to the Perl program is extracted from Perlxs (2018). The produced Perl function will accept that its first contention is an object pointer. The object pointer will be put away in a variable called THIS. The object are written in C++ with the new( ) work and are executed by Perl with the sv_setref_pv( ) macro. The object by Perl can be dealt with by a typemap. For example, if the C++ code shown below

```
"       class colour {
        public:
        colour();
            ~colour();
        int blue();
        void set_blue( int );
        private:
        int c_blue;
    };
"
```

is to be linked to the Perl program using THIS.

```
" RETVAL = THIS->blue();
THIS->set_blue( val );"
```

So that the link will look like below

```
"

int
colour::blue( val = NO_INIT )
  int val
  PROTOTYPE $;$
CODE:
  if (items > 1)
    THIS->set_blue( val );
    RETVAL = THIS->blue();
  OUTPUT:
    RETVAL

"
```

References

Capterra. (2017). *Environmental software*. https://www.capterra.com/environmental-software/. Accessed on January 4th, 2018.

Chandra, K. B., & Chinmayee, R. (2012). Incorporating hidden Markov model into anomaly detection technique for network intrusion detection. *International Journal of Computer Applications, 53*(11), 42–47.

ConnectUS (2018). 7 Main Advantages and Disadvantages of Open Source Software, https://connectusfund.org/7-main-advantages-and-disadvantages-of-open-source-software. Accessed November 12th, 2018.

Emetere, M. E., & Sanni, E. S. (2015). A Review on the comparative roles of mathematical softwares. *Global Journal of Pure and Applied Mathematics, 11*(6), 4937–4948.

Flory, W. J. (2018). *3 tips for organizing your open source project's workflow on GitHub.* https://opensource.com/article/18/4/keep-your-project-organized-git-repo. Accessed August 25th, 2018.

Higgs, P. (2016). *The disadvantages of open source.* https://www.gaiaresources.com.au/opensource/. Accessed January 4th, 2018.

Ibrahim, H. (2010). https://www.linuxfoundation.org/resources/open-source-guides/starting-opensource-project/. Accessed August 25th, 2018.

Igor, B. (2017). *Top 15 Python libraries for data science in 2017.* https://medium.com/activewizards-machine-learning-company/top-15-python-libraries-for-data-science-in-in-2017-ab61b4f9b4a7. Accessed on January 4th, 2018.

István, F., Ágnes, H., Zahari, Z. (2013). *Advanced numerical methods for complex environmental models: Needs and availability* (p. 201). https://doi.org/10.2174/97816080577881130101; eISBN: 978-1-60805-778-8, 2013. ISBN: 978-1-60805-777-1.

Korolkovas, A. (2016). *Entangled polymer flows at interfaces.* A thesis submitted to University of Upsala.

Matthias, S. (2013). *Four types of open source communities.* https://opensource.com/business/13/6/four-types-organizational-structures-within-open-source-communities. Accessed August 25th, 2018.

NASA. (2018). *Open-source software project.* https://code.nasa.gov/. Accessed January 4th, 2018.

NOAA. (2018). *Numerical weather prediction*. https://www.ncdc.noaa.gov/data-access/model-data/model-datasets/numerical-weather-prediction. Accessed January 9th, 2018.

OSI. (2018). https://opensource.org. Accessed August 25th, 2018.

Patwardhan, M. (2016). *Assessing the impact of usability design features of an mHealth app on clinical protocol compliance using a mixed methods approach*. Arizona State University: ProQuest Dissertations Publishing; 2016. https://repository.asu.edu/attachments/172769/content/Patwardhan_asu_0010N_16210.pdf

Perlxs, (2018). http://perldoc.perl.org/perlxs.html#Using-XS-With-C%2b%2b. Accessed August 27th, 2018.

Seher, R. (2017). *Combining open source software licenses—The final chapter*. http://blog.thehyve.nl/blog/open-source-software-licenses-3. Accessed August 25th, 2018.

Tom, A. (2004). *How to misunderstand open source software development*. http://www.consultingtimes.com/ossdev.html. Accessed August 25th, 2018.

Upasani, O.S. (2016). Advantages and limitations of open source software for library management system functions: The experience of libraries in India. *The Serials Librarian, 71*(2), 121–130. https://doi.org/10.1080/0361526X.2016.1201786.

Välimäki, M. (2005). *The rise of open source licensing: A challenge to the use of intellectual property in the software industry* (*Ph.D. thesis*). Helsinki University of Technology. Retrieved 2015-12-30.

Wilbanks, J. (2013). *Understanding open science*. http://fastercures.tumblr.com/post/56790751132/understanding-open-science. Accessed August 24th, 2018.

Chapter 3
An Overview of Theoretical Dynamics of Air Pollution

Air pollution maybe outdoor or indoor. In this chapter the scope is on the outdoor air pollution. Outdoor air pollution can be natural or man-made/artificial. Both kind of pollution sources have its hazards to life-forms. Most natural air pollutions are connected to gas emission from volcanic eruption, dust storm (Sahara Desert in the West Africa region), particulate matter (PM) carried by wind, industries (see Figs. 3.1 and 3.2) etc. Therefore, modelling outdoor air pollution is somewhat difficult because it is not a closed system. A closed system is a system where the component parameters can be controlled or determined at each step of the experiment. In a close system, physical laws are mostly obeyed since the input parameter and the output parameter in the exchanging environment is known.

Fig. 3.1 Air pollution from industry

© Springer Nature Switzerland AG 2019
M. E. Emetere, *Environmental Modeling Using Satellite Imaging and Dataset Re-processing*, Studies in Big Data 54,
https://doi.org/10.1007/978-3-030-13405-1_3

Fig. 3.2 Air pollution from volcanic eruption

Outdoor air pollution is an open system. Hence many factors like source-emissions, wind transport, meteorology, relative pollution speed, atmospheric concentrations, rate of particulate-coagulation, deposition, lifetime of particulates, type of anthropogenic pollution are considered in its model formulation. Also, many mathematical assumptions are propounded so that some physical laws can be applied to obtain approximate results. Since outdoor air pollutions are also dynamic, approximate results cannot be relied-upon for a long time. This reason is also affecting the reliability of measuring instruments whose design is based on theoretical model. For example, Bock et al. (2008) reported that some sondes have design errors. Researcher have clamoured that a continual review of accepted models is essential to minimize error margin—among other shortcoming (Emetere and Akinyemi 2017; Wilson et al. 2014; Nuret et al. 2008).

In has been reported that some outdoor air pollutants have adverse effect on the health of children. Prominent among the outdoor pollutants are particulate matter, ozone, nitrogen oxides, carbon monoxide and sulfur dioxide. Results from satellite imagery reveal that children in the West African region highly exposed to outdoor air pollution. Hence, the reason for the increase in lung defects in children between the ages of 10 and 18 years is known (Gauderman et al. 2004). A rough scientific investigation (by the author) to support the assertions of Gauderman et al. (2004) was carried out using spatial map of aerosol loading and proven mathematical model over Damaturu in northern Nigeria. With an assumption that the respiratory component of the human lungs as presented in Fig. 3.3, it is logical to understand the diversion of particulate matter in the human body.

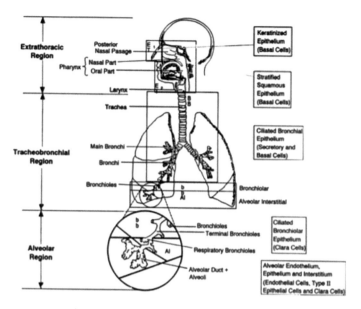

Fig. 3.3 The respiratory component of human being

The entrance of outdoor pollutant into the lungs has a long journey to travel into other parts of the body. During the pollutants travel time, the particulates have chances of reacting chemically with components from tissues and organs. For children, the survival chances are very slim. In Damaturu–Nigeria, the aerosol loading in the region is high. From the spatial map, the sources of the outdoor pollution are known (Fig. 3.4). Figure 3.5 shows the deposition efficiency of outdoor pollutants into the human lungs.

The deposition efficiency is higher in April and May of every year. The radius of the pollutants is also shown in Fig. 3.6. Hence, the concept of outdoor pollution is of interest.

In the above scenario, the risk ratio for mortality was modelled by O'Neill et al. (2003) as:

$$\log RR_{ij} = \beta_0 + \beta_1 \times personal\ risk\ factors_i$$
$$+ \beta_2 \times O_{3j} + \beta_3 \times \left(O_{3i} \times O_{3j}\right)$$
$$+ \beta_4 \times O_{3i} \times SEP_i + V_j + U_j \times O_{3j}$$

where I represents an individual in the jth area; V_j is the random area effect that captures variation in risk, not documented as personal risk factors within the geographic area; and U_j is the variation in the slope of O_3, not explained by the individual level interaction terms within the geographic area. β_2 measures the effect of O_3 exposure within the geographic area and β_3 is the effect of the difference from that area wide exposure for the ith individual. β_4 is the effect modification of individual SEP on

Fig. 3.4   Aerosol loading map over Damaturu

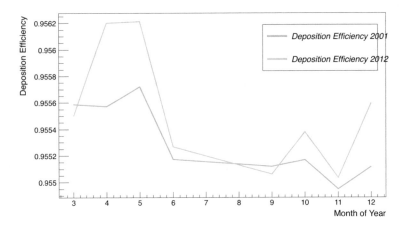

Fig. 3.5   Deposition efficiency over Damaturu

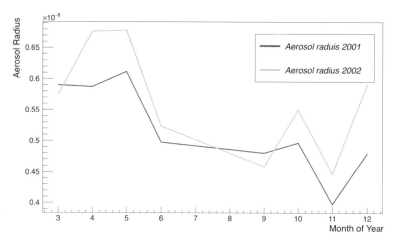

Fig. 3.6 Monthly display of aerosol radius

Table 3.1 Estimated emission of primary pollutants in the USA 1987 (AQPC)

Source	CO	Particles	H/C's	NO_x	SO_x	Total	%
Transport	40.7	1.4	6.0	8.4	0.9	57.4	44.9
Stationary fuel combustion	7.2	1.8	2.3	10.3	16.4	38.0	29.7
Industrial processes	4.7	2.5	8.3	0.6	3.1	19.2	15.0
Solid waste disposal	1.7	0.3	0.6	0.1	0	2.6	2.0
Miscellaneous	7.1	1.0	2.4	0.1	0	10.6	8.3
Total	61.4	7.0	19.6	19.5	20.4	127.8	
%	48.0	5.5	15.3	15.3	15.9		100

the response to O_3. With the above formula, it is easy to estimate the risk factor (as presented in Table 3.1) of emission of primary pollutants in the USA.

In general, outdoor air pollution models are designed under the auspices of pollution-source models (Emetere and Akinyemi 2013; Boubel et al. 1994), urban pollution models (Lindén et al. 2012), sea and land pollution models (Norris et al. 2013), meteorological models (Emetere et al. 2015a, b; Gazala et al. 2006), dispersion models (Emetere et al. 2017a, b; Holmes et al. 2006), photochemical model (Vladutescu et al. 2013), particulate-transport model (Emetere et al. 2017c), deposition/coagulation model (McKibbin 2008), numerical models (Emetere 2016b), statistical model (Emetere 2016a), analytical models (McKibbin 2008), computational models (Emetere et al. 2015a, b), geostatistical model (Gualtieri and Tartaglia 1998) etc. Some of the models are adopted when investigating general properties of outdoor air pollutants e.g. numerical models, statistical model, analytical models, computational models. The general properties of outdoor air pollutants include optical, chemical, transport, physical and radioactive properties.

Macdonald (2003) categorized air pollution techniques into scale. The lowest in the scale is the gross screening models. The gross screening model is an air pollution technique that requires hand-held calculator, monograph, or spreadsheet. Simple equations used in this type of model are formula for estimating worst case mean concentrations downwind of a point source (Hanna et al. 1996). The formula is given as:

$$C_{wc} = \frac{10^9 Q}{U H_{wc} W_{wc}}$$

where Q is the source strength or emission rate of gas or particulate (kg/s), C_{wc} is the worst case concentration ($\mu g/m^3$), U is the worst case wind speed at height z = 10 m, usually 1 m/s, W_{wc} is the worst case cloud width (m) (usually assume W = 0.1x, where x is distance from the source), H_{wc} is the worst case cloud depth (usually assume H = 50 m in worst case). Another equation used in the gross screening models is equation to calculate the concentration of species (in ppm)

$$C_i = \frac{8.314 T V}{p M_i}$$

where V is the concentration in μgm^{-3}, T is the temperature, M is the molecular weight of species, p is the pressure (in pascal).

The next air pollution technique in the low scale is the advanced models. It requires a desktop PC or workstation were large dataset for meteorology and emissions are considered. Most advanced models are purely computational model with lots of simulations without mathematical representative. However, there are some intrigues in such model by the inclusion of multiple source types (point, area and volume), complex terrain, flow around buildings and layered atmospheric structure.

The next air pollution up the scale is the specialized models. This type of model is often used for predicting air pollution dispersion in larger scale. Types of specialized models include Agency Regulatory Model (AERMOD), CALPUFF, VISTAS Version 6 model, Community Multiscale Air Quality (CMAQ), Comprehensive Air quality Model (CAMX), Urban Airshed Model (UAM), CALGRID, CALMET, Mesoscale Model version 5 (MM5), Regional Atmospheric Modeling System (RAMS), Kinematic Simulation Particle Model (KSP), MONTECARLO, Chemical Mass Balance Model (CMB) etc. Other alternative models are ADAM, ADMS 3, AFTOX, DEGADIS, HGSYSTEM, HOTMAC, RAPTAD, HYROAD, ISC3, OBODM, PANACHE, PLUVUEII, SCIPUFF, SDM, SLAB, AERSCREEN, CTSCREEN, SCREEN3, TSCREEN, VALLEY, COMPLEX1, RTDM3.2, VIS-CREEN, REMSAD, UAM-V, HYSPLIT, PUFF-PLUME, Puff model, ADMS-3, ADMS-URBAN, ADMS-Roads, ADMS-Screen, GASTAR, NAME, UDM, AEROPOL, Airviro, Airviro Grid, Airviro Heavy Gas, ATSTEP, AUSTAL2000, BUO-FMI, CAR-FMI, CAR, DIPCOT, DISPERSION21, DISPLAY-2, EK100W, FARM, FLEXPART, GRAL, HAVAR etc. The specialized models will be discussed in details in succeeding sections.

Air pollution cannot be discussed without focusing on its measurement techniques. The measurement techniques engender the degree of reliance or confidence on the dataset that emanates from the process. It also helps to understanding measuring errors and how to prevent it. The most common air pollution measuring technique is the sampling methods. This method requires a control which is the analyzed ambient air without modifying its chemical composition. Any sample of air pollutants that will be measured is shielded from rainfall and dust that can largely influence measurement data. There are two types of sampling methods namely passive sampling method and active sampling method. The application of any of these sampling types depends on the discretion of the scientist.

Passive sampling is the technique used when estimating spatial gradients around compliance monitors, surveying additional pollutants, identifying hotspots, and estimating individual exposure while the active sampling is used when there is the need for accurate and precise flow rates and sample durations.

The second type of air pollution measuring technique is the particle scattering technique. This technique requires the need of nephelometer with a PM2.5 inlets and a smart heater to remove moisture under high relative humidity (e.g. >65%). This technique is not too common. The third air pollution measuring technique is the Gas chromatography. This method is common in the measurement of greenhouse gases like CO, CO_2, CH_4, SF_6 and N_2O.

The fourth air pollution measuring technique is the UV fluorescence. UV fluorescence is similar to UV spectrophotometry. The ultraviolet radiation that is emitted from the sample is measured by knowing the wavelength of the fluorescence component's. In order to avoid interference with the existing radiation, measurement is carried out in a large angle from the radiation direction. The fifth air pollution measuring technique is the Chemo-luminescence method. This method is used to measure the photochemical NO, NO_2 and NO_x concentrations.

The sixth air pollution measuring technique is the Spectrophotometry. Spectrophotometry is based on the absorption of radiation by gases at different wavelengths. For example, if light with known intensity is passed through the sample, the emerging radiation's intensity determines the composition of the gas. The Beer–Bougert–Lambert-law is used in this type of technique.

The seventh air pollution measuring technique is remote sensing. Remote sensing measurements are used to satisfy the following needs: estimation of ozone concentration in the stratosphere; high spatial and temporal resolution air quality data to estimate fluxes; high spatial resolution air quality data to provide initial fields for transport-exchange models; vertical profile measurements; air quality information of hardly accessible areas; and measurement of airborne pollutants.

The eighth air pollution measuring technique is rainwater analysis. Rainwater analysis consists of two main parts namely pH measurement and composition measurement. PH measurement is used to estimate acidic wet deposition often referred to as acidic rain while compositional measurement is used to estimate wet deposition of soluble gases and minerals. The accepted levels of major air pollutants is shown in Table 3.2.

Table 3.2 Accepted levels of major air pollutants (ACTRAC 1998)

Air pollutant	Acceptable level
CO	1 h ave. 30 ppm (60 ppm detrimental) 8 h ave. 9 ppm (20 ppm detrimental) 1 h alert level 150 ppm
NO_2	1 h ave. 0.12 ppm (0.25 ppm detrimental) 8 h ave. 0.06 ppm (0.15 ppm detrimental) 1 h alert level 0.50 ppm 1 year 0.03 ppm
NH_3	Ground level conc. 0.83 ppm (0.6 mg/m^3)
HNO_3	Ground level conc. 0.067 ppm (0.17 mg/m^3)
SO_2	1 h ave. 0.20 ppm (0.34 ppm detrimental) 8 h ave. 0.06 ppm (0.11 ppm detrimental) 1 day ave. 0.08 ppm 1 year ave. 0.02 ppm 1 h alert level 0.50 ppm
H_2S	Ground level conc. 0.0001 ppm (0.00014 mg/m^3)
Photochemical oxidants (as O_3)	1 h ave. 0.10 ppm (0.15 ppm detrimental) 4 h ave. 0.08 ppm (0.15 ppm detrimental) 8 h ave. 0.05 ppm (0.08 ppm detrimental) 1 h alert level 0.25 ppm
Respirable particles	24 h ave. 120 mg/m^3 (240 mg/m^3 detrimental) 1 year ave. 40 mg/m^3 (80 mg/m^3 detrimental)
PM10 respirable	1 day ave. 50 μg/m
Atmospheric lead	3 month ave. 1.0 μg/m^3 1 year ave. 0.50 μg/m^3
Benzo[α]pyrene	1 year ave. 5.0 ng/m
Benzene	1 year ave. 10.0 ng/m
Fluorine	Ground level conc. 0.033 ppm (0.067 mg/m^3)

The measuring techniques have been highlighted above. However, the measuring instruments or device are quite important because some of the equipment have reported to have biases that could compromise the measuring processes. These biases are sometimes attached to the climate system or location the device operates. For example, Emetere (2016a) reported measuring device over West Africa must be reconfigured to avoid breakdown of the instrument or adulteration of the measuring processes. Therefore, the selection of locations for monitoring equipment is as important as the measuring device. When these two conditions (location and equipment) are met, then the measuring techniques will be effective. For example, in the sampling methods (whether passive or active sampling), an appropriate position is required before doing the measurement. So what are these appropriate positions? An appropriate position is a position where the best measurement can be obtained. For example, pollution sources where the pollutants are dispersed or deposited. Appropriate positioning is the first factor to site selection. The other factors include purpose of monitoring, number and type of instruments required, and duration of measurements. In addition to site selection, site monitoring should possess some guideline as stipulated in AS2922 (Australia standard 1987). Monitoring sites should be easily accessible to avoid data loss. Data loss is currently a concurrent challenge professional in the environmental disciplines encounter. Many attempts to mitigate the rate of data loss have been suggested in the past i.e. optimize location choice by accessing past experiments conducted over the region and avoiding local interferences. So what happens if the air quality of the location is not known? Hence, the author suggests that state-of-the-art equipment should be fabricated for higher accuracy and resolution. Meteorological Monitoring has been suggested to be a crucial factor to site monitoring equipment. Meteorology helps us to understand the influence of atmospheric forces to determine air quality at any particular place and time. Through meteorology, we understand that pollutants travel distances and deposits at different points, thereby making the pollution at the source equally important as particle-deposited locations. The atmospheric forces include wind speed, wind magnitude, surface temperature, moisture in the atmosphere, photochemical reactions, precipitation rate and rainfall. Every location has its wind pattern. Within the wind pattern the wind direction, wind speed, wind magnitude and wind stagnation can be known. Wind stagnation is a condition where the wind sensor or anemometers detects no wind magnitude. In such situation, if air pollutants are emitted into the atmosphere, the aerosol loading will be very high with a short period of time. It is then important to know how to install the anemometers. The standard exposure of anemometers over open terrain is 10 m above the ground. The anemometer should not be installed at locations where airflow is obstructed by buildings, hills or tall trees. Hence, the anemometer should be installed fifty meters away and fifteen meters higher than obstacles.

The surface temperature influences the updraft of moisture which in turn reacts with the air pollutants in the atmosphere. Also, the surface temperature is affected by the sizes of pollutants because the coagulating agent (moisture) may be missing at high surface temperature. In this case, smaller pollutants are present in the atmosphere.

The ambient air monitoring is an important factor for site selection. This factor helps the scientist to understand the type of air pollutants presents in the location. Hence, for almost every type of air pollutant, there are several different acceptable methods for analyzing them. There are factors that drive the use of ambient air monitoring. These factors include cost, meteorology of the location, number of data points required, purpose for which the data are being used, data storage methods, time interval required between data points, devices power requirements, type of air pollutant, and the environment in which the monitoring equipment is being placed.

With the advent of portable and affordable air pollution devices, some of the old theories and procedure is gradually becoming obsolete. Some of the equipment have sensitivity that is ten times higher than the orthodox equipment (e.g. Spirometers, displacement bottles, soap bubble meter, mercury sealed piston, roots meters, wet test gas meters, dry test gas meters etc.). Hence, the criteria for ambient air monitoring may not be needed. However, like the orthodox equipment, new air monitoring equipment requires regular calibration.

Some of the new portable equipment are Air Quality Temperature Humidity Meter PCE-HT110 (for measuring humidity with a resolution of 0.1 °C/0.18 °F, 0.1% RH), Air Quality Particle Counting Meter PCE-RCM 10 (for measuring particulate matter (PM) concentrations in the air with a resolution of -20 to $+70$ °C/-4 to $+158$ °F and 0–100% RH), Air Quality VOC Meter PCE-VOC 1 (for measuring volatile organic compound (VOC) with resolution of TVOC range of 0.00–9.99 ppm and HCHO range of 0.00–5.00 ppm), Air Quality Meter PCE-RCM 11 (for measuring fine dust, temperature and relative humidity with resolution of Particle size channels of PM 2.5/PM 10, TVOC range of 0.00–9.99 mg/m^3 temperature range of -20 to 70 °C/-4 to 158 °F, humidity range of 0–100% RH and formaldehyde range of 0.00–5.00 mg/m^3), Air Quality Particle Counting Meter PCE-RCM 12 (for measuring CO_2, fine dust, temperature and relative humidity with resolution of Particle size of PM 2.5/PM 10, temperature range of -20 to $+70$ °C/-4 to $+158$ °F, Humidity measuring range of 0–100% RH, Formaldehyde measuring range of 0.00–5.00 mg/m^3, and CO_2 measuring range of 0–9999 ppm), Air Quality Particle Counting Meter PCE-MPC 10 (for measuring particulate matter (PM) concentrations in the air with resolution of particle sizes of 2.5 and 10 μm, temperature measuring range of 0 to $+50$ °C/$+32$ to $+122$ °F, and humidity measuring range: 0–100% RH), Air Quality Carbon Dioxide Meter PCE-WMM 50 (for measuring carbon dioxide (CO_2) with resolution of CO_2 gas measurement range of 0–50,000 ppm, temperature measurement range of 0 to $+45$ °C/$+32$ to $+113$ °F), Air Quality Temperature Humidity Meter PCE-G1 (for measuring relative humidity (% RH) and temperature with a resolution of humidity ranges of 10–95% RH/0 to $+60$ °C, accuracy relative humidity/temperature of ±2% RH/1 °C), Air Quality Particle Counting Meter PCE-PCO 1 (for measuring the concentration of particles in the air with a resolution to detect particle sizes of 0.3, 0.5, 1.0, 2.5, 5.0, and 10 μm), Air Quality Carbon Dioxide Meter CDL 210 (for measuring carbon dioxide (CO_2) levels as well as relative humidity (% RH) and ambient temperature with a resolution of 1 ppm of CO_2, measurement range of 0–2000 ppm of CO_2, and accuracy of ±50 ppm $\pm$ 5% of reading of CO_2), Air Quality Meter Gasman-H$_2$S "Hydrogen Sulfide" (for measuring hydrogen sulfide gas with a resolution of 0–100 ppm), Air Quality Meter Gasman-O$_2$ "Oxygen" (for detecting the existence of dangerous gas concentrations and shows on screen the value of the gas measurement with a resolution of 0–25% v/v), Air Quality Meter Gasman-FL "Flammable Gases" (for detecting dangerous gas concentrations and shows on screen the value of the gas measurement with a resolution of 0–100% LEL), Air Quality Meter Gasman-NO$_2$ "Nitrogen Dioxide" (for measuring nitrogen dioxide with a resolution of 0–10 ppm), Air Quality Meter Gasman-NH$_3$ "Ammonia Gas" (for measuring ammonia with a resolution of 0–100 ppm), Air Quality Meter Gasman-O$_3$ "Ozone" (for measuring ozone in the atmosphere with a resolution of 0–1 ppm), Air Quality Meter Gasman-PH$_3$ "Phosphine" (for measuring phosphine with a resolution of 0–5 ppm), Air Quality Formaldehyde Gas Meter HFX205 (for measuring formaldehyde gas with a resolution of 0–10 ppm) etc.

3.1 Limitation to Outdoor Air Pollution Model

Pollution-source model may be divided into four types (i.e. according to its shape) namely, point source, line source, area source and volume source. These sources may be stationary or mobile. The most popular pollution-source model is the plume model. The plume model maybe subdivided into Gaussian-plume model, looping plume model, Buoyant plume model, Dense gas plume model, conning plume model, basic plume model, stack plume model, rising-plume model, Passive or neutral plume model, general plume dispersion model etc. From the theoretical point of view, the plume models have their limitation or drawback. First, plume model (Gaussian-plume model) assume that pollutants travel in a straight line i.e. from its source to its destination or receptor that maybe very far from each other. Technically, this means that the plume model cannot be used to analyze causality or risk effects. Second, plume models are disadvantaged when applied to low wind speed situation. This is due to the inverse wind speed dependence of the steady-state plume equation. Third, plume model does not account for turbulent wind situation (GSL 2018). Fourth, the plume models use steady state approximations and do not take into account the time required for pollutants to travel from source to the receptor or destination.

The dispersion model is one of the versed models in air pollution. There are lots of theoretical inputs and assumptions in the dispersion model to satisfy basic physical laws. There are five types of dispersion models namely Box model, Gaussian model (Gaussian plume and puff model), Lagrangian model, Eulerian model, Dense gas model, puff model etc. Though the dispersion model can be seen as an improvement on the limitations of the plume model, but it has its limitations also. First, the dispersion model requires much dependency on meteorological input data to validate, estimate or test the accuracy of the model. Second, there are lots of scenario to justify its accuracy. For example, a dispersion model should be incorporate the elemental realities of boundary-layer meteorology, atmospheric chemistry, wind dynamics, atmospheric turbulence, mesoscale meteorology, particulate retention and particle dynamics.

Emetere et al. (2017a, b, c) solicited that particulate retention should be considered differently. Most literature on dispersion model assumes that particulates have continuous motion throughout an event. This assumption is the third limitation of the dispersion model. In reality, particulates may be static during coagulation or wind re-orientation. Also, particulates may maintain a virtuous travel path. At this state, particulate retention in the atmosphere takes place. The sustenance of this particulate retention depends on the life-time of the particulates and external atmospheric influences. The physics proposed theory have not been validated on a large scale.

Fourth, the validation dispersion model may be compromised by regional meteorology. Hence, working on a global scale may require much computing resources (e.g. high-specification computers, processing time, disk space etc.) to assimilate the influences of regional meteorology. This idea, bring us back on the appropriate open-source library to analyze big-data. The fifth limitation of the dispersion model is the assumption that mass diffusion can be ignored due to the large mass transfer of particulate bulk motion in x, y, z-directions. This mean the dispersion model may not accurately describe pollution dynamics in risk assessment. The slightest mass diffusion may change the rate of air-pollution mixing.

Jacobson et al. (2007) clearly highlighted the limitations in the urban pollution model (e.g. global-through urban air pollution-weather-climate model). The vertical velocities in urban pollution model are calculated under a hydrostatic rather than non-hydrostatic assumption. In the opinion of the author, this limitation may not be alleviated because of the different boundary-layer that exist over urban centers. The limitation of the geostatistical model is its dependence on the density of the monitoring network (Jerret et al. 2005). The limitation of the meteorological models is the recurrent need to have a large dataset to draw reliable conclusion (Emetere et al. 2015a, b). The other limitations of the meteorological models are inappropriateness of the data and lack of control over data quality. The limitations of the photochemical models include the design of photochemical reactors, disposal or recycling process, production of sludge, pH range limitations (Bahnemann and Robertson 2015), systematic spatial and temporal biases (Crooks and Ozkaynak 2014).

The numerical environmental model is becoming the most popular method adopted by scientist, engineers and environmentalist to show the sensitivity of their model. More so, most numerical problems are solved using computational techniques. In other words, numerical environmental modelling seems easier based on trivial assessment of the process. However, the most technical aspect of numerical environmental modelling is the formation of realistic assumptions. The first limitation is that some assumptions are not absolute necessary due to breakthroughs in open-source numerical library. In the case of a complex system e.g. large-scale air pollution model, the limitation may be the approach the researcher adopts in solving the mathematical problems (i.e. systems of partial or ordinary differential equations). In recent time, this limitation may not be noticeable because most numerical environmental model has friendly or interactive user-interface that allows the upload of input variables. Hence, users of such software are not privy to the mathematical solutions in the software or model. The next sub-section illustrates the many solutions obtained from a large-scale air pollution model.

3.2 Dynamics of Large-Scale Air Pollution

In the last section, it was emphasized that one of the limitation of the numerical model is the researcher approach for solving a complex case e.g. large-scale air pollution. In this section, the focus is to highlight the many large-scale air pollution formulation and the different solution that was obtained. This would help beginner to appreciate the modelers' effort in developing a working model. Also, it will help beginner to be curious on the basic formulation of whatsoever model or software they have adopted for research, study or analysis. It is expedient for reader to know the composition of air pollution in large scale. The European Environmental Agency gave the dataset that relates to 28 countries in Europe—in 1994. The summary of the air pollution

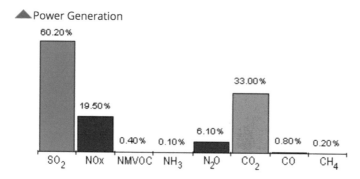

Fig. 3.7 Air pollution composition from power generation (EEA 2016)

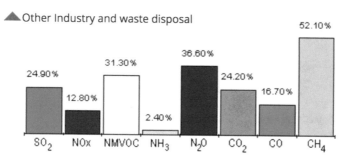

Fig. 3.8 Air pollution composition from industry (EEA 2016)

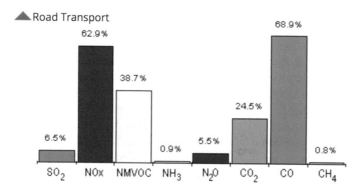

Fig. 3.9 Air pollution composition from road transport (EEA 2016)

from different sources were illustrated in Figs. 3.7, 3.8, 3.9, 3.10 and 3.11. The air pollutant that was examined was SO_2, NO_x, NMVOC, NH_3, N_2O, CO_2, CO and CH_4. The sources of pollution that was identified were from power generation, industry, road transport, domestic sources and agriculture (EEA 2016).

Carbon monoxide (CO) is regarded as one of the major constituents of air pollution globally. The volume of occupancy of carbon monoxide in the Earth's atmosphere

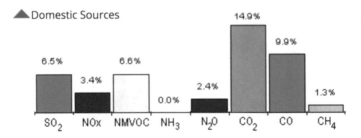

Fig. 3.10 Air pollution composition from domestic sources (EEA 2016)

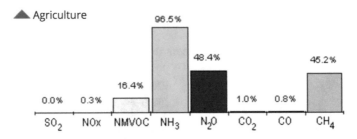

Fig. 3.11 Air pollution composition from agriculture (EEA 2016)

is around 530 million ton (about 0.00001%), with an average residence time of 36–100 days. The sources of CO include, vehicular emission, volcanic eruptions, industry, bush burning, power generation (through coal burning) etc. The annual global emissions are estimated to be 3×10^9 to 6.4×10^{11} ton per year for natural and 2.75×10^8 ton per year for man-made sources.

Carbon dioxide is produced almost like the CO. It is removed from the atmosphere by plants during photosynthesis. Also, CO_2 are removed from the atmosphere via its solubility in water. Hence, the ocean is regarded as major sink of CO_2. Base values of CO_2 have reportedly increased to about 25%—since 1850. SO_2 is a major air pollutant. It is estimated that 65 million ton of SO_2 per year enter the atmosphere via anthropogenic emission, volcanic eruption etc. Nitric oxide (NO) is listed as air pollutants. The natural emissions of NO are estimated to be approximately 5×10^8 ton per year. The sources of NO emission is anthropogenic sources. Concentrations background concentrations of NO and NO_2 are approximately 0.5 and 1 ppb respectively.

Ammonia is one of the common air pollutants that are found in abundance in the atmosphere. The EEA charts presented in Figs. 3.7, 3.8, 3.9, 3.10 and 3.11 is a clear evidence of its abundance in the atmosphere. Approximately 4×10^6 ton are emitted per year on a worldwide basis. Most ammonia emission emanates from biological decomposition. Ammonia in the atmosphere is oxidized in the atmosphere in a series of chemical reactions to produce nitrates. Methane is regarded as air pollutants that are generated from human and agricultural waste. Methane has also been recognized as one of the trace gases that may have significant effect on global climate through the greenhouse effect. Methane is by far the most abundant hydrocarbon in the atmosphere.

Volatile organic compounds (VOCs) are airborne gaseous or vaporous substances of organic origin. Examples of VOCs include hydrocarbons, alcohols, aldehydes and organic chemicals. VOCs can become airborne through evaporation (from chemicals, cleaning detergent, solvents, paints, varnishes, adhesives) or emission (anthropogenic sources like building site, bush burning, domestic waste etc.). These impurities can be slowly released from the product's surface into the air. Hence, VOC is both an indoor and outdoor air pollutants.

Khaled et al. (2014) considered the modelling of atmospheric dispersion with dry deposition. The research focused on the formulation of three dimensional advection-diffusion equations to simulate the dispersion of pollutants in the planetary boundary layer. The advection-diffusion equation that describes pollutants dispersion in a turbulent medium was given:

$$\frac{\partial C}{\partial t} + u\frac{\partial C}{\partial x} + v\frac{\partial C}{\partial y} + w\frac{\partial C}{\partial z} = \frac{\partial}{\partial x}\left(K_x\frac{\partial C}{\partial x}\right) + \frac{\partial}{\partial y}\left(K_y\frac{\partial C}{\partial y}\right)$$
$$+ \frac{\partial}{\partial z}\left(K_z\frac{\partial C}{\partial z}\right) + S + R \tag{3.1}$$

u, v and w are the constant wind speed in the x, y and z directions respectively. K_x, K_y and K_z are the eddy diffusion coefficients along the x, y and z directions. C is the mean concentration of the pollutants/contaminants, S is source/Sink term, t is the time. Equation (3.1) was solved analytically by considering two cases under the steady state condition i.e. the downwind distance x and the vertical height z. Hence the solutions for the two scenarios were given as:

$$C(x, y, z) = \frac{Q\lambda}{u\sigma_y\sqrt{2\pi h}\,J_1\left(2\lambda\sqrt{h}\right)} J_0\left(2\lambda\sqrt{z}\right)\exp\left(-\frac{ku*x}{u}\lambda^2\right)\exp\left(-\frac{y^2}{2\sigma_y^2}\right)$$
$$\tag{3.2}$$

J_0 is the Bessel function of the first kind of order 0, J_1 is the Bessel function of the second kind of order 1, k is the von-Karman's constant, u* is the friction velocity, h is the height of the mixing layer, Q is the emission rate and δ_y is the Dirac. delta function, and λ is a separation constant

$$C(x, y, z) = \frac{Q}{uh\sigma_y\sqrt{2\pi}}\exp\left(-\frac{v_d}{ku_*}\left(\frac{z-h}{x} - \frac{v_d}{2u}\right)\right)\exp\left(-\frac{y^2}{2\sigma_y^2}\right) \tag{3.3}$$

v_d is the deposition velocity. From the above, it is noted that the difference between both scenarios (i.e. Eqs. 3.2 and 3.3) are the main factor that makes pollutant matter transport not uniform at both directions.

$$f = \frac{\lambda\sqrt{h}}{J_1\left(2\lambda\sqrt{h}\right)} J_0\left(2\lambda\sqrt{z}\right) \exp\left(-\frac{ku*x}{u}\lambda^2\right)\left(\frac{v_d}{ku_*}\left(\frac{z-h}{x} - \frac{v_d}{2u}\right)\right) \quad (3.4)$$

From the above, it can be inferred that the pollutants transport is far greater on the horizontal direction by a factor of 'f'. It can be roughly inferred that pollutants would travel about 100,000 times horizontally than vertical. Looking at same Eq. (3.1), Sutton (1932) gave the solution as:

$$C(x, y, z) = \frac{Q}{2\pi_y u \sigma_z} \exp\left(-\frac{y^2}{2\sigma_y^2}\right) \times \left\{\exp\left(-\frac{(z-H)^2}{2\sigma_z^2}\right) + \exp\left(-\frac{(z+H)^2}{2\sigma_z^2}\right)\right\}$$
$$(3.5)$$

The comparison between the solutions (Eqs. 3.3 and 3.5) of two different authors at different age corroborate the views of the author in the last section. Assumptions are indeed a sort of limitation—if it does not describe an event. Most modeler see assumptions as a leeway to solving mathematical equations. However, caution on its misuse is advised.

Tirabassi et al. (2010) worked on the comparison between Non-Gaussian Puff Model and a model based on a time-dependent solution of advection-diffusion equation. The advection-diffusion equation that was considered is given as:

$$\frac{\partial C}{\partial t} + u\frac{\partial C}{\partial x} + v\frac{\partial C}{\partial y} = \frac{\partial}{\partial z}\left(K_z\frac{\partial C}{\partial z}\right) + K_h\left(\frac{1}{d^2}\frac{\partial^2 C}{\partial x^2} + \frac{\partial^2 C}{\partial y^2}\right)$$
$$+ \delta(t)\delta(x)\delta(y)\delta(z-1) \quad (3.6)$$

δ is the wind velocity vector, K_h is eddy diffusivity. Other parameter remains the same as earlier defined. The focus of the researcher was to simulate the pollutant dispersion in the Planetary Boundary Layer (PBL). The solution of the advection-diffusion equation was obtained using the truncated Gram-Charlier expansion (type A) of the concentration field and finite set equations for the corresponding moments. The solution is given as:

$$C(x, z, t) = \sum_{i=1}^{k} w_i\left(\frac{P_i}{t}\right) \sum_{j=1}^{M} w_j\left(\frac{P_j}{x}\right)\left[A_n exp - \left(\sqrt{\frac{P_i}{tK_n} + \frac{P_j u_n}{xK_n}}\right)\right.$$
$$- B_n \exp(z - H_s)\left(\sqrt{\frac{P_i}{tK_n} + \frac{P_j u_n}{xK_n}}\right)$$
$$+ \frac{1}{2}\frac{Q}{\sqrt{\frac{P_i}{t} + \frac{P_j u_n}{x}}K_n}\left[\begin{array}{l}\exp -(z - H_s)\left(\sqrt{\frac{P_i}{tK_n} + \frac{P_j u_n}{xK_n}}\right) \\ \left. - \exp(z - H_s)\left(\sqrt{\frac{P_i}{tK_n} + \frac{P_j u_n}{xK_n}}\right)\right]\right] \cdot H(z - H_s)$$
$$(3.7)$$

$H(z - H_s)$ is the Heaviside function, w_i and w_j are the weights, P_i and P_j are the roots of the Gaussian quadrature scheme, K_n is the average eddy current, u_n is the average wind speed, A_n and B_n is the coefficients. The authors claimed that the solution accurately describe the meteorological data (Copenhagen data set). The number of assumptions in the article gives much concern on the model relevance to describe larger data set.

Macdonald (2003) worked on the theory of air dispersion modelling for large scale. The advection-diffusion equation that was used is shown below:

$$\frac{\partial C}{\partial t} + u\frac{\partial C}{\partial x} = \frac{\partial}{\partial y}\left(K_y\frac{\partial C}{\partial y}\right) + \frac{\partial}{\partial z}\left(K_z\frac{\partial C}{\partial z}\right) + S \tag{3.8}$$

The analytical solution of the above equation was given as

$$C(x, y, z) = \frac{Q}{2\pi u \sigma_y \sigma_z}\exp\left(-\frac{y^2}{2\sigma_y^2}\right) \times \left\{\exp\left(-\frac{(z - H)^2}{2\sigma_z^2}\right) + \exp\left(-\frac{(z + H)^2}{2\sigma_z^2}\right)\right\} \tag{3.9}$$

All parameters remain as defined earlier. For emphasis, consider the root advection-diffusion Eqs. (3.1) and (3.8). It is clear that the solutions of Macdonald (2003), Sutton (1932) and Khaled et al. (2014) are product of various assumptions. Considering the solutions of Macdonald (2003) and Sutton (1932) in Eqs. (3.5) and (3.8), it can be inferred that despite their assumptions, the factor that separates both solutions is given as σ_y. This immediately raises the question on the accuracy of one of the solutions. Again, this example also corroborates the fact that making 'assumptions' might actually be a form of drawback in applying the numerical methods to explaining large scale air pollution.

Walcek (2004) investigated on the wind shear influence on the Gaussian dispersion/plume model. The steady-state advection-diffusion equation that was used is given as:

$$u\frac{\partial C}{\partial x} + v\frac{\partial C}{\partial y} = K_h\frac{\partial^2 C}{\partial y^2} + K_z\frac{\partial^2 C}{\partial yz^2} \tag{3.10}$$

The solution of Eq. (3.10) is given as:

$$C(x, y, z) = \frac{Q}{2\pi u \sigma_y \sigma_z \sqrt{1 + s^2/12}}\exp\left[\frac{-y^2}{2\sigma_y^2\left(1 + \frac{s^2}{12}\right)}\right.$$
$$\left. + \frac{-z^2\left(1 + \frac{s^2}{3}\right)}{2\sigma_z^2\left(1 + \frac{s^2}{12}\right)} + \frac{yz}{2\sigma_y\sigma_z}\left(\frac{s}{1 + \frac{s^2}{12}}\right)\right] \tag{3.11}$$

s is the non-dimensional shear factor given as

$$s = \frac{\partial v}{\partial z} \frac{x}{u} \sqrt{\frac{K_z}{K_h}} \tag{3.12}$$

The comparison between the parent Eqs. (3.8) and (3.10) and the modalities of their respective solution (Eqs. 3.9 and 3.11) show the following:

i.   Both researchers worked on a steady-state advection equation. This means that Macdonald (2003) made $\frac{\partial C}{\partial t} = 0$.

ii. The introduction of $v\frac{\partial C}{\partial y}$ changed the nomenclature of the solution such that only one exponential term can be seen in Eq. (3.11)

iii. The so called final solution i.e. Eq. (3.11) cannot be applied to live dataset without approximating the parameter defining 's'.

Emetere (2016b) developed an advection-diffusion equation as shown below:

$$\frac{\partial C}{\partial t} + V_x \frac{\partial C}{\partial x} - V_z \frac{\partial C}{\partial z} - V_y \frac{\partial C}{\partial y} = \frac{\partial}{\partial z}\left(K_z \frac{\partial C}{\partial z}\right) + \frac{\partial}{\partial y}\left(K_y \frac{\partial C}{\partial y}\right)$$
$$+ \frac{\partial}{\partial z}\left(K_{z2} \frac{\partial C}{\partial z}\right) + \frac{\partial}{\partial y}\left(K_{y2} \frac{\partial C}{\partial y}\right) - P + S \tag{3.13}$$

P is the air upthrust. The researcher considered a steady state advection with emphasis on the topography of the study area. Equation (3.13) was reduced to:

$$V_z \frac{\partial C}{\partial z} = \left(K_z \frac{\partial^2 C}{\partial z^2}\right) + \left(K_y \frac{\partial^2 C}{\partial y^2}\right) + \left(K_z \frac{\partial^2 C}{\partial x^2}\right) \tag{3.14}$$

Considering the wind turbulence within the study area and its effect on the dispersion of pollutants, the solution of Eq. (3.14) was given as:

$$C(x, y, z) = a^2 b \cos\left(\frac{n\pi}{2}x\right) \cos\left(\frac{n\pi}{2}y\right) \exp\left(-\frac{V_z}{k_z}z\right) \tag{3.15}$$

Like the solution presented by Walcek (2004), Eq. (3.15) also had an exponential term. Hence, the nomenclature of the solution changes with respect to the nature of the advection-diffusion equation and type of assumption made by the modeler.

The intent of this section is not to play-down on erstwhile research done. Also, it is not to show the superiority of a model over the other. However, the section only lay foundation on why the succeeding chapters were considered. Computational image/data processing has advanced tremendously such that satellite imageries could be re-processed to get more interesting information. Also, satellite dataset could be processed at a reduced computational time. This means that the calibration, validation, verification and sensitivity analysis of the computational technique can be done simultaneously. This invention is truly novel such that source images could be faulted, corrected or modified to comprehend all the vital information of the imagery.

Large scale air pollution measurement is significant as the model that has been considered above. The first measuring instrument that measures at large scale is satellite observation. The satellite observation that has been highlighted in this paragraph are National Aeronautics and Space Administration (NASA) satellites. There are several satellite that are not linked to NASA in the globe. The listing of the following satellite observation was for illustration purpose. Aqua was launched in 2002. It is used to observe interactions in oceans, land, atmosphere, and biosphere. Tropical Rainfall Measuring Mission (TRMM) was launched in 1997 to study global precipitation. Jason-2 was launched in 2008 to study ocean surface height. Jason-3 was launched in 2016 as an improvement on the Jason-3 satellite. The Cloud-Aerosol Lidar and Infrared Pathfinder Satellite Observation (CALIPSO) was launched in 2006. It is used to study clouds and aerosols so as to comprehend compositional changes in the atmosphere. AURA was launched in 2004. It is used to study the composition, chemistry, and dynamics of the atmosphere. Laser Geometric Environmental Observation Survey (LAGEOS) 1 was launched in 1976 for geodynamical studies. LAGEOS 2 was launched in 1992 as an improvement to LAGEOS 2. CloudSat was launched in 2006. It is used to study clouds with the aim of determining rainfall, snowfall, and moisture content of clouds. SEASTAR (SEAWIFS) was launched in 1997 to study the color of earth's oceans. Global Precipitation Measurement was launched in 2014 to study global precipitation. Geostationary Operational Environmental Satellite GOES I-M was launched in 2001 to study and forecast weather. Gravity Recovery and Climate Experiment Follow-On (GRACE-FO) was launched in 2018 to study gravity and climate. Landsat-7 was launched in 1999 to study earth's coastal areas with global coverage on a seasonal basis. Landsat 8 was launched in 2013 as an improvement to Landsat-7. Proba-V was launched in 2013 to study the earth's vegetation. ICESat was launched in 2003 to study the size and thickness of earth's ice sheets. QuikSCAT was launched in 1997 to study weather using microwaves. Solar Radiation and Climate Experiment (SORCE) was launched in 2003 to study the earth's absorption of radiation energy. Deep Space Climate Observatory was launched in 2015 to monitor the Sun-lit side of Earth from the L1 Lagrange point. TERRA was launched in 1999 to study the interaction between solar radiation and the state of the atmosphere, land, and oceans.

Also, there are some ground instruments for measuring air pollution in large scale. The trending equipment is the weather station. It is a facility with instruments and equipment to make observations of atmospheric conditions in order to provide information to make weather forecasts and to study the weather and climate. AQS 1 Construction Air Quality Monitor (Fig. 3.12) is a type of workstation that carries the following functions: Measure the key pollutants building locales; Screen particulate matter (TSP, PM10 and PM2.5) and nitrogen dioxide (NO_2) at the same time; Exceptionally precise part-per-billion (ppb) recognition of NO_2; Incredible relationship to US EPA reference analyzers; Can be aligned in fields for most extreme traceability; PM bay warmer to make up for moistness impacts; Measures and reports information in 1-min interims with user selectable averaging; On board stockpiling for more than 5 long periods of information; Tough weatherproof nook with sunlight based protecting for extremely sweltering atmospheres; Snappy set up and migration in less

Fig. 3.12 AQS 1
construction air quality
monitor

than 10 min; Email/SMS alerts and FTP data transfer; displays ecological sensors (discretionary).

Gill weather stations and wind sensors (Fig. 3.13) has the ability to monitor wind speeds from as low as 0.1 m/s with no mechanical latency, so the data provided is extremely reliable and high quality. It has no mechanical moving parts that require frequent replacement, in this manner operational expenses are commonly low.

Locally made solar-powered weather station works on modular design of ATmega328P microcontroller and uses the ESP8266 Serial-to-Wifi chip to upload the weather data to the server. The weather station receives power from a 20 W solar panel that charges a 12 V lead acid battery using a custom built solar charge controller. The sensors in the device include Sparkfun weather meters (wind wane, anemometer and rain gauge); Adafruit AM2315 (temperature and humidity sensor); Adafruit BMP280 (pressure); and DFROBOT PM 2.5 laser dust sensor (air quality) (Fig. 3.14).

EE-WMS-AQM automated weather station (Fig. 3.15) is a device that can be used for large scale air pollution monitoring. The versatility of the automated weather station is evidenced in the ability of user to add, remove, or substitute sensors or other peripherals as your data measurement and monitoring needs change.

It has the following functionality: a convenient filter tool provided for the separation of aerosol droplets and condensate from gases; unattended weather recording at remote and exposed sites; in-built system software with data file management; large

Fig. 3.13 Gill weather
stations and wind sensors

Fig. 3.14 Locally made
solar-powered weather
station

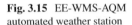

Fig. 3.15 EE-WMS-AQM
automated weather station

storage capacity, and they store data permanently; GPRS modem communications that is secure.

Vaisala WXT-520 automatic weather station (Fig. 3.16) is a veritable device for monitoring large scale air pollution because it has the standard parameters of a weather station packed into a single, compact case. Parameters include air temperature, relative humidity, barometric pressure, wind speed and direction (using a top quality sensor), plus a revolutionary electronic precipitation gauge that can not only detect rain, but also its intensity. It is very easy to install, and no moving parts makes it extremely easy to maintain.

AQS 1 Urban Air Quality Monitor (Fig. 3.17) is a reliable device for large scale air pollution monitoring. It has the following functionalities: measures the key pollutants of concern in the urban environment; monitor particulate matter (PM), nitrogen dioxide (NO_2) and ozone (O_3) at the same time; proven 1 part-per-billion (ppb) detection of O_3 and NO_2; highly selective O_3 and NO_2 measurements, no cross-interference; very high correlation to US EPA reference analyzers; can be field calibrated for maximum traceability; inlet heater to compensate for humidity effects; measures and reports data in 1 min intervals with user selectable averaging; and Email/SMS alerts and FTP data export.

HYDRA Dual Sampler (Fig. 3.18), is a reliable measuring device that is used to measure large scale air pollution. The instrument can work with any sampling inlet (PM10, PM2.5, PM1) within the operating flow rate range 0.8–2.5 m^3/h, on two distinct independent channels.

Fig. 3.16 Vaisala WXT-520 automatic weather station

Fig. 3.17 AQS 1 urban air
quality monitor

Fig. 3.18 HYDRA dual
sampler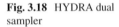

Condensation Particle Counter—CPC3775 (Fig. 3.19) is a reliable device to measure large scale air pollution. Its design is almost like the Hydra dual sampler. It detects airborne particles down to 4 nm. It provides highly accurate measurements over a wide concentration range from 0 to 107 particles/cm^3. This CPC is quite versatile and well-suited for a broad range of applications.

AQS 1 Remediation Air Quality Monitor (Fig. 3.20) is a versatile device for monitoring large scale air pollution. It has the following functionality: measure the key pollutants found at site remediation and landfill projects; monitor particulate matter (TSP, PM10 and PM2.5), volatile organic compounds (VOC) and optional nitrogen dioxide (NO$_2$) simultaneously; highly accurate part-per-billion (ppb) detection of VOCs and NO$_2$; excellent correlation to US EPA reference analyzers; PM inlet heater to compensate for humidity effects; measures and reports data in 1-min intervals with user selectable averaging; rugged weatherproof enclosure with solar shielding for very hot climates; Email/SMS alerts and FTP data export.

AQS 1 Smog Monitor (Fig. 3.21) is a versatile device for monitoring large scale air pollution. It has the following functionality: measure the key indicators pollutants in photochemical smog; continuous real-time measurement of PM2.5 and ozone (O$_3$); 1 part-per-billion (ppb) detection of ozone; no cross-sensitivity to nitrogen dioxide (NO$_2$); very high correlation to US EPA reference analyzers; inlet heater to compensate for humidity effects; measures and reports data in 1 min intervals with user selectable averaging; rugged weatherproof enclosure with solar shielding for very hot climates; Email/SMS alerts and FTP data export.

Fig. 3.19 Condensation
particle counter—CPC3775

Fig. 3.20 AQS 1
remediation air quality
monitor

Fig. 3.21 AQS 1 smog
monitor

Air Pollution Monitoring station (Fig. 3.22) that is manufactured by Arrow Instruments Calibration is very useful in the industry by providing a high-quality array of Air Pollution Monitoring. It measures contaminants in the air, such as carbon monoxide (CO), nitrogen dioxide (NO_2), ozone (O_3), particulate matter (PM2.5 and PM10), sulfur dioxide (SO_2), and hydrogen sulfide (H_2S). The device use sophisticated technology that makes it easier to use.

OPSA-150 Organic Pollutant Monitor (Fig. 3.23) is a device designed for large scale pollution measurement. It is a new organic pollutant monitor that uses HORIBA's proprietary Rotary Cell Length Modulation. The unit can be used as an organic pollutant monitor at drainage systems for determining compliance with COD monitoring regulations, for monitoring quality of water measuring levels of organic matter at water supply intakes, and as an organic monitor on process lines (phenol meter).

Ambient Air PM2.5 Samplers (Fig. 3.24) is an improvement upon APM 550 EL, APM 550 Mini, APM 550 MFC and APM 550. It is a tool for measuring large scale air pollution. Its special features include: PM10 and PM2.5 impactors of sampler based on designs standardized by US EPA; brushless, Oil-free, light weight pump practically requires no maintenance; filter holder designed for any 47 mm diameter filter media; critical Orifice maintains constant sampling rate of 1 m^3/h; provision to attach gaseous sampler; and compact cabinet design for easy portability.

Ambient Air PM10 Samplers (Fig. 3.25) is large scale air pollution measuring device. Its modular design, APM 460 NL enables it to be easily paired with a gaseous

Fig. 3.22 Air pollution
monitoring station (arrow
instruments calibration)

Fig. 3.23 OPSA-150
organic pollutant monitor

Fig. 3.24 Ambient air
PM2.5 samplers

sampling attachment (for monitoring SO_2, NO_x, NH_3, Ozone etc.) as gaseous sampling requires only a few LPM of air flow. This is possible through an attachable subsidiary unit APM 411 or the more modern APM 411 TE. Kindly refer to respective brochures of APM 411 and APM 411TE for details. Its special features include: essentially diminished noise when contrasted with most PM 10 instruments in the market; blower with in-manufactured warm cut-off that makes the device not to need a stabilizer; brushless blower diminishes gear downtime and upkeep exertion; lockable best cover; improved bureau plan which is more tough and sturdy with SS equipment; electromagnetic Interference (EMI) to TVs completely eliminated.

Airpointer's modular (Fig. 3.26) design comprising a base unit, analyzing modules and sensor modules allows for a configuration according to different application requirements. It is used for monitoring airborne pollutants (SO_2, NO_2/NO_X, CO, O_3, and PM) classified as relevant by the EU, the WHO, the US-EPA and further responsible organizations all over the world. It also has a fast optical system or an approved PM analyzer for monitoring PM. It has an integrated data management system records for monitoring data of the airpointer's own analysis modules as well as various external third-party sensors. It has internal web server enables data retrieval by using any Internet connection.

Fig. 3.25 Ambient air PM10 samplers

Fig. 3.26 Airpointer's modular

References

ACTRAC. (1998). *Environmental control—Unit 2 air pollution*, Cert. In Chem. Plant Skills Resource.

Bahnemann, D. W., & Robertson, P. K. J. (2015). Environmental photochemistry part III. In *The handbook of environmental chemistry* (p. 307). New York: Springer. ISBN-13: 978-3662467947, ISBN-10: 3662467941.

Bock, O., Bouin, M. -N., Doerflinger, E., Collard, P., Masson, F., Meynadier, R., et al. (2008). The West African monsoon observed with ground based GPS receivers during AMMA. *Journal of Geophysics Research, 113*, D21105.

Boubel, et al. (1994). *Fundamentals of air pollution* (3rd ed). Academic Press.

Crooks, J., & Ozkaynak, H. (2014). Simultaneous statistical bias correction of multiple PM2.5 species from a regional photochemical grid model. *Atmospheric Environment, 95*, 126–141.

EEA. (2016). *Sources of air pollution*. https://www.eea.europa.eu/publications/2599XXX/page010.html#note. Accessed August 25, 2018.

Emetere, M. E. (2016a). Statistical examination of the aerosols loading over Mubi-Nigeria: The satellite observation analysis. *Geographica Panonica, 20*(1), 42–50.

Emetere, M. E. (2016b). *Numerical modelling of West Africa regional scale aerosol dispersion*. Thesis submitted to Covenant University.

Emetere, M. E., & Akinyemi, M. L. (2013). Modeling of generic air pollution dispersion analysis from cement factory. *Analele Universitatii din Oradea-Seria Geografie, 231123-628*, 181–189.

Emetere, M. E., & Akinyemi, M. L. (2017). Documentation of atmospheric constants over Niamey, Niger: A theoretical aid for measuring instruments. *Meteorological Applications, 24*(2), 260–267.

Emetere, M. E., Akinyemi, M. L., & Akinojo, O. (2015a). A novel technique for estimating aerosol optical thickness trends using meteorological parameters. *2015 PIAMSEE: AIP Conference Proceedings, 1705*(1), 020037.

Emetere, M. E., Akinyemi, M. L., & Uno, U. E. (2015b). Computational analysis of aerosol dispersion trends from cement factory. In *IEEE Proceedings 2015 International Conference on Space Science & Communication* (pp. 288–291).

Emetere, M. E., Sanni, S. E., Emetere, J. M., & Uno, U. E. (2017a). Thermal infrared remote sensing of hydrocarbon in Lagos-Southern Nigeria: Application of the thermographic model. *International Geomate Journal, 13*(39), 33–45.

Emetere, M. E., Esisio, F., & Oladapo, F. (2017b). Satellite observation analysis of aerosols loading effect over Monrovia-Liberia. *Journal of Physics: Conference Series, 852*(1), art. no. 012009. https://doi.org/10.1088/1742-6596/852/1/012009.

Emetere, M. E., Sanni, S. E., & Tunji-Olayeni, P. (2017c). Atmospheric configurations of aerosols loading and retention over Bolgatanga-Ghana. *Journal of Physics: Conference Series, 852*(1), art. no. 012007. https://doi.org/10.1088/1742-6596/852/1/012007.

Gauderman, W. J., Avol, E., Gilliland, F., Vora, H., Thomas, D., Berhane, K. R., et al. (2004). The Effect of Air Pollution on Lung Development from 10 to 18 Years of Age, J Med, *351*, 1057.

Gazala, H., Venkataraman, C., Isabelle, C., Ramachandran, S., Olivier, B., & Shekar, M. R. (2006). Seasonal and interannual variability in absorbing aerosols over India derived from TOMS: Relationship to regional meteorology and emissions. *Atmospheric Environment, 40*, 1909–1921.

GSL. (2018). *Getting started in atmospheric dispersion modelling—An introduction*. https://guides.co/g/atmospheric-dispersion-modelling-an-introduction/24917. Accessed January 7, 2018.

Gualtieri, G., & Tartaglia, M. (1998). Predicting urban traffic air pollution: A GIS framework. *Transportation Research, D3*(5), 329–336.

Hanna, S. R., Drivas, P. J., & Chang, J. C. (1996). Guidelines for Use of Vapor Cloud Dispersion Models. AIChE/CCPS, 345 East 47th St., New York, NY 10017, 285 pp.

Holmes, N. S., & Morawska, L. (2006). A review of dispersion modelling and its application to the dispersion of particles: an overview of different dispersion models available. *Atmospheric environment, 40*, 5902–5928.

Jacobson, M. Z., Kaufman, Y. J., & Rudich, Y. (2007). Examining feedbacks of aerosols to urban climate with a model that treats 3-D clouds with aerosol inclusions. *Journal of Geophysical Research, 112,* D24205. https://doi.org/10.1029/2007JD008922.

Jerret, M., Arain, A., Pavlos, K., Bernardo, B., Dimitri, P., Talar, S., et al. (2005). A review and evaluation of intraurban air pollution exposure model. *Journal of Exposure Analysis and Environmental Epidemiology, 15,* 185–204.

Khaled, S. M. E., Soad, M. E., & Maha, S. E. (2014). Modelling of atmospheric dispersion with dry deposition: An application on a research reactor. *Revista Brasileira de Meteorologia, 29*(3), 331–337.

Lindén, J., Thorsson, S., Boman, R., & Holmer, B. (2012). *Urban climate and air pollution in Ouagadougou, Burkina Faso: An overview of results from five field studies* (pp. 1–88). University of Gothenburg. http://hdl.handle.net/2077/34289.

Macdonald, R. (2003). Theory and objectives of air dispersion modelling. *Modelling Air Emissions for Compliance, Wind Engineering, MME, 474A,* 1–27.

McKibbin, R. (2008). Mathematical modeling of aerosol transport and deposition: Analytic formulae for fast computation. In *Proceedings of International Congress on Environmental Modeling* (pp 1420–1430).

Norris, S. J., Ian, M. B., & Dominic, J. S. (2013). A wave roughness Reynolds number parameterization of the sea spray source flux. *Geophysical Research Letters, 40,* 4415–4419.

Nuret, M., Lafore, J. P., Bock, O., Guichard, F., Agustı̀-Panareda, A., Ngamini, J. B., et al. (2008). Correction of humidity bias for Vaısala RS80 sondes during AMMA 2006 observing period. *Journal of Atmospheric Oceanic Technology, 25,* 2152–2158.

O'Neill, M. S., Jerrett, M., Kawachi, I., Levy, J. I., Cohen, A. J., Gouveia, N., et al. (2003). Health, wealth, and air pollution: Advancing theory and methods. *Environmental Health Perspectives, 111,* 1861–1870.

Standards Australia. (1987). *AS2922—A guide for the siting of sampling units.*

Sutton, O. G. (1932). A theory of eddy diffusion in the atmosphere. *Proceedings of the Royal Society London A, 135,* 143–165.

Tirabassi, T., Moreira, D. M., Vilhena, M. T., & da Costa, C. P. (2010). Comparison between non-gaussian puff model and a model based on a time-dependent solution of advection-diffusion equation. *Journal of Environmental Protection, 1,* 172–178.

Vladutescu, D. V., Bomidi, L. M., Barry, M. G., Qi, Z., & Shan, Z. (2013). Aerosol transport and source attribution using sun photometers, models and in-situ chemical composition measurements. *IEEE Transactions on Geoscience and Remote Sensing, 51*(7), 3803–3811.

Walcek, C. J. (2004). *A Gaussian dispersion/plume model explicitly accounting for wind shear.* https://ams.confex.com/ams/pdfpapers/79742.pdf. Accessed January 9, 2018.

Wilson, R., Luce, H., Hashiguchi, H., Nishi, N., & Yabuki, Y. (2014). Energetics of persistent turbulent layers underneath mid-level clouds estimated from concurrent radar and radiosonde data. *Journal of Atmospheric and Solar-Terrestrial Physics, 118*(A), 78–89.

Chapter 4
Image Re-processing of Satellite Imageries

In this chapter, the focus is re-processing satellite imageries. Satellite imagery are images of Earth or other planets collected by Imaging satellites. The quality of satellite imagery is judged by its resolution. The resolution associated to satellite imagery are namely spatial, spectral, temporal, geometric and radiometric (Campbell 2002). The spatial resolution is the pixel size of an image that represents the size of the surface area. Spectral resolution is the discrete segmentation of the electromagnetic spectrum that describes the number of intervals and interval size of the sensor wavelength. Temporal resolution describes the time-duration between imagery collection periods for a given surface location. Geometric resolution describes the ability of the satellite sensors to effectively image a portion of the Earth's surface in a single pixel. Radiometric resolution describes the ability of imaging system to record many levels of brightness in bit-depths of 8-bit (0–255), 11-bit (0–2047), 12-bit (0–4095) or 16-bit (0–65,535).

The quality of the resolution depends on the satellite view and the imaging camera. For example, 'Landsat 7' 15 m satellite imagery is not as sharp as either the 'Spot' 10 m or 'IRS-1C' 5 m satellite imagery but it has large area coverage (Landinfo 2018). GeoEye-1 has an orbit altitude of about 770 km/478 Miles and is capable of producing imagery with a ground sampling distance of 46 cm, meaning it can detect objects of that diameter or greater (GeoEye 2018). NASA (2015) revealed the high-tech of the Earth Polychromatic Imaging Camera (EPIC) used for Deep Space Climate Observatory (DSCOVR) satellite. EPIC takes a series of 10 images using different narrowband filters from ultraviolet to near infrared to produce a variety of science products. One of the novel features of EPIC is that it has a very fast image processing time.

In this chapter, the satellite imagery that was re-processed is the GES-DISC Interactive Online Visualization ANd aNalysis Infrastructure (Giovanni). Giovanni users have access to satellite imagery from multiple remote sites. Its image processing platform supports multiple data formats namely Hierarchical Data Format (HDF), HDF-EOS, network Common Data Form (netCDF), GRIdded Binary (GRIB), GIF, and binary. Also, Giovanni have multiple plot types e.g. area, time, Hovmoller, and

© Springer Nature Switzerland AG 2019
M. E. Emetere, *Environmental Modeling Using Satellite Imaging
and Dataset Re-processing*, Studies in Big Data 54,
https://doi.org/10.1007/978-3-030-13405-1_4

image animation. Giovanni users have free-access to data on atmospheric chemistry, atmospheric temperature, water vapor and clouds, atmospheric aerosols, precipitation, ocean chlorophyll and surface temperature (Acker and Leptoukh 2007). Its primary satellite imagery consists of global gridded data sets with reduced spatial resolution. It is based upon this fact, the need to re-process the satellite imagery using open source software (CERN-ROOT) and library (OpenCV C++) was borne. This gesture is meant to unravel deeper information of dataset through the adoption of modern computational techniques.

4.1   Image Processing

The satellite imagery that was used covers West Africa and some few countries around it. The scope of the satellite imagery is aerosols optical depth for January–December, 2007 and January–December, 2013. The raw satellite image was converted from '.gif' to 'png' to ensure an enhanced image. The image properties of the raw satellite data and the converted data are shown in Table 4.1. It is observed that the new images are better based on the dpi (Dots Per Inch) width, dpi height, sample per pixel, pixel width and pixel height.

The raw satellite image of aerosol optical depth pixel count is shown in Fig. 4.1.

The pixel of Fig. 4.1 was re-processed in 8 bits image. The y-pixel was potted against the x-pixel to obtain the original image in another format whose information is inverse to the original satellite imagery (Fig. 4.1). In Fig. 4.1, the AOD pixel counts is highest at the south of Chad, Senegal and Mali; north of Cameroun, Benin, Togo, Ghana, Cote d'Ivoire, Sierra Leone, Gambia; north-central of Nigeria and Guinea.

The redefinition of the satellite image shows the following

Table 4.1 Image properties

Number	Parameters	Old image	New image
1	pixelWidth	700	1242
2	pixelHeight	500	646
3	typeIdentifier	com.compuserve.gif	public.png
4	format	gif	png
5	dpiWidth	72,000	144,000
6	dpiHeight	72,000	144,000
7	samplesPerPixel	3	4
8	bitsPerSample	8	8
9	hasAlpha	no	yes
10	space	RGB	RGB
11	profile	sRGB IEC61966-2.1	Color LCD

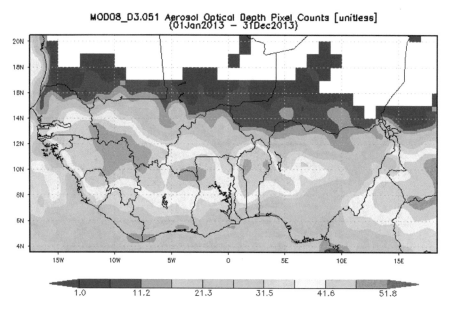

Fig. 4.1 Satellite image AOD pixel counts-Jan. to Dec., 2013

i.   The inactive part (white region) of Fig. 4.1 was reconstructed Fig. 4.2.
ii. A clear spot in the north-central of Nigeria that may typify a special feature that require more investigation.
iii. Evidence of massive aerosol retention across latitude $7°–13°N$ which queries the authenticity of the major wind that affect West Africa region between latitudes $9°$ and $20°N$ (Rafferty 2010).
iv. The wind dynamics i.e. southwesterly that blows during warmer months and northeasterly that blows during cooler months have low impact on aerosols distribution within latitude $7°–13°N$.
v.   There is a significant aerosol path (red lines) that continuously connects from land to sea and vice versa.

The contour detection algorithm was used to determine contours not visible to the eye. The circles are used to measure the width of the contour i.e. the diameter of the circle. The blue arrow shows the insignificant circle because the contour inside the circle is a regional boundary. The red arrow shows the relevant circle which contains contours that cannot be spotted by the human eye (Fig. 4.3). 40% of the circles actually points to the useful contours while about 60% only relates to boundary lines. The sizes of the circle can be measured and the given locations within the circle can be investigated for more information.

The 3D image (Fig. 4.4) shows the relation between the derived matrix of the image. The 70–150 pixel (along Y-axis) corresponds to locations under intense influence of the Sahara dust. The 0–50 pixels (along Y-axis) show locations on Fig. 4.1 that have mechanisms for dousing the influence of Sahara dust via wind convection

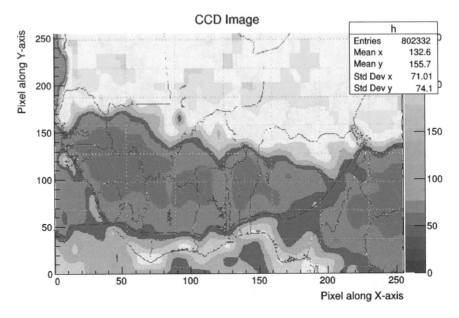

Fig. 4.2 x and y pixel redefinition of the satellite image

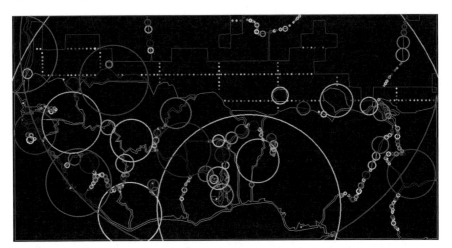

Fig. 4.3 Contour detection of satellite image

or rainfall patterns (Emetere et al. 2017a, b; Emetere 2017). One of the importance of Fig. 4.4 is the characterization of the vertical profile. For example, within 0–70 pixel, the vertical or z-axis of Fig. 4.4 show that at the lower troposphere, there may be high aerosol particulates. Hence, the background aerosol content (BAC) over a region maybe known. While some location between latitude 14° and 20°N have decreased aerosol particulates as it disperses from the troposphere into the stratosphere, loca-

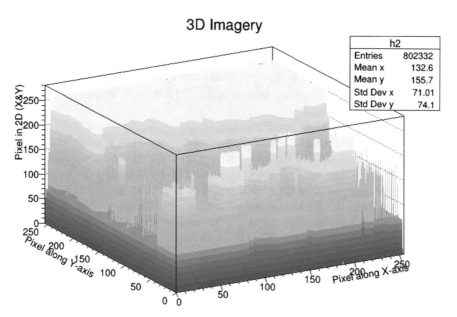

Fig. 4.4 3D setting of satellite image

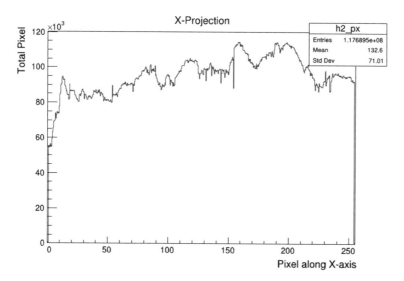

Fig. 4.5 X-projection of satellite image

tions between latitude 7° and 13°N have increased aerosol particulates as it disperses from the troposphere into the stratosphere.

Figure 4.5 show the projection along the x axis of Fig. 4.1. Projection along x or y takes its root from the mathematical Eq. (4.1) below:

$$\vec{X}_k^h(s) = \begin{pmatrix} \vec{X}_k^0 \\ 1 \end{pmatrix} + s \begin{pmatrix} \vec{t} \\ 0 \end{pmatrix} \tag{4.1}$$

here $\vec{t}$ is tangent direction, s is the free parameter for points along the line, $\vec{X}_k^0$ is the arbitrary points on the image coordinates, and $\vec{X}_k^h(s)$ is the parallel lines in the image coordinates.

Normally, the projection along an image axis should be represented by parallel lines, however, there are situation of distortions due to the properties of the image. Hence, a parallel line may be reconstructed as a Gaussian line or multiple sinusoidal line. Image projection along axes can be influenced by the intrinsic and extrinsic calibration of the camera, feature of the image (i.e. radiometry, reflection and colour), digital image formation, and image noise. Image reconstruction of the projection may be influenced by the algebraic nature of the parallel line (Eq. 4.2), frequency domain (Eq. 4.3) and filtered back projection (Eq. 4.4).

$$p_m = \sum_{n=1}^{N} w_{m,n} f_n = w_m f \tag{4.2}$$

$w_{m,n}$ is the weight, p_m is the stripe sum, and f_n is the vector of the 'n' unknown image-pixel.

$$F(u, v) = \int_{-\infty}^{\infty} \int_{-\infty}^{\infty} f(x, y) \exp j(ux + vy) dx dy \tag{4.3}$$

$F(u, v)$ is the 2D image spectrum, $f(x, y)$ is the image pixel in the x and y axes, u and v are the positions in the x and y axes respectively.

$$b(x, y) = \int_{0}^{\pi} P_\theta(\tau) d\theta \tag{4.4}$$

$b(x, y)$ is the back-projection operator, τ is the discretization of projections, P_θ angle dependent stripe sum. The type of image projection discussed here is the filtered-back projection.

The projection along the x axis relates to the x-component of the satellite image in magnitude of pixels along the x-axis of Fig. 4.1. For example, on the pixel 150 and 200, the x-component of the total pixels is highest over Nigeria, Chad etc. Also, two points (165 and 190) was maximum which clearly shows that there are higher aerosols dispersion along these points. Another way of describing or reporting projections along either x or y axis, is by the number, shape and magnitude of clear peaks. The peaks in Fig. 4.5 are on the pixel-10 (Gambia and Senegal), pixel-85 (Cote d'Ivoire and southern Mali), pixel-120 (Ghana and Burkina Faso), pixel-160 (Nigeria) and pixel-200 (Nigeria and northern Cameroun).

Also, considering the Y-Projection of Fig. 4.1 i.e. the magnitude of pixels along the y-axis, it can be observed that the pixel reaches its peak values above 170,000 and

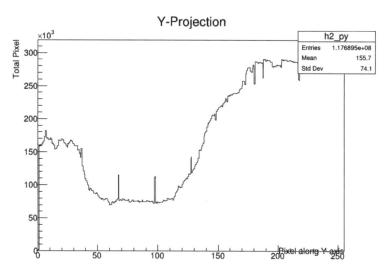

Fig. 4.6 Y-projection of satellite image

290,000. The peak was found at >40 pixel (southern parts of Liberia, Cote d'Ivoire, Ghana, Togo, Benin and Nigeria) and >180 pixel (Mauritania, Mali and Niger). From Fig. 4.1, the region where the peak is found have low aerosol count pixel. The two minor peaks found at the trough of the shape i.e. Fig. 4.6 are at pixel-68 (northern Sierra Leone, northern Liberia, southern Guinea, southern Ghana, southern Togo, southern Benin, central Nigeria and northern Cameroun) and pixel-99 (central Guinea, northern Cote d'Ivoire, northern Ghana, northern Togo, northern Benin and Central Nigeria).

The spectrum analysis of the satellite image (Fig. 4.1) is shown in Fig. 4.7. The spectrum analysis shows the vertical profile and turbulence of aerosols. The aerosols activities i.e. dispersion, distribution and loading are illustrated by the color-mix represented. For example, within 30–110 pixel on the x-axis arrow shows region of highest aerosol count pixel-mostly due to Sahara dust and anthropogenic pollution in the given locations. The vertical profiles (as seen on the z-axis) show that aerosol forms layers within the atmosphere. The blue cross-sectional area shows region of low aerosol distribution.

Aerosol extinction optical depth at 388 nm was obtained from OMI/Aura level-2 near UV Aerosol data product (OMAERUV). OMAERUV measures aerosols in five atmospheric levels namely, Aerosol Layer Height (ALH), Normalized Radiance (NR), Lambert equivalent Reflectivity (LER), Surface Albedo (SA), and Imaginary Component of Refractive Index (ICRI). Aerosol Extinction optical depth is a measure of the annihilation of the solar beam by dust and haze.

In Fig. 4.8, the high AEOD can be found both on land (Niger, Chad, Mali, Mauritania and fragment-parts in southern Nigeria) and sea (coastal part of West Africa). The high values in Niger, Chad, Mali and Mauritania is due to the Sahara dust (Lindén

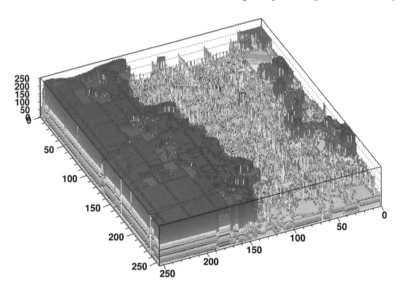

Fig. 4.7 The spectrum analysis of the satellite image

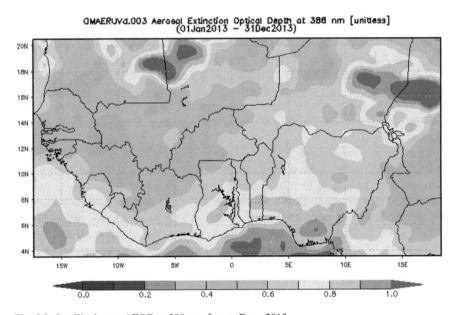

Fig. 4.8 Satellite image AEOD at 388 nm, Jan. to Dec., 2013

et al. 2012). The high value of AEOD at the fragment-parts in southern Nigeria is due to gas flaring (Omotosho et al. 2015).

Recall that it was acknowledged that the colour bar in Fig. 4.2 was inversely proportional to the satellite image. In like manner it was observed that Fig. 4.9

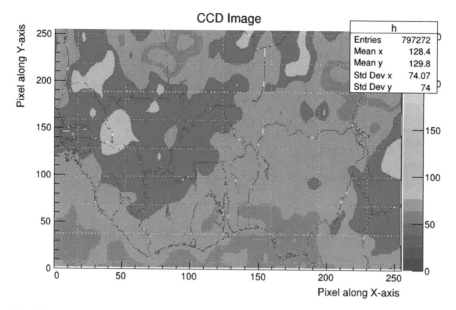

Fig. 4.9 x and y pixel redefinition of the satellite image

contradicts the satellite image in Fig. 4.8. The dark blue patches depict the highest AEOD which was not the case in Fig. 4.9. The affected areas were central Nigeria, northern Ghana, northern Cote d'Ivoire, Burkina Faso, Mali, Mauritania, Senegal, Gambia, Guinea and northern Sierra Leone. The validation of this site has been reported (Emetere 2016a, b; Emetere et al. 2017a, b, c; Emetere and Akinyemi 2017). The most interesting part of the image processing is that it shows the highest and lowest colour representation of the aerosol properties investigated.

Figure 4.10 show the circles surrounding some interesting features that cannot be seen in Figs. 4.8 and 4.9. For example, the biggest circle over Nigeria show region that have been confirmed by several research to have intense air pollution due to gas flaring (Emetere et al. 2015a, b; Omotosho et al. 2017). All the big circles show similar deductions that would be substantiated in the next chapter.

Figure 4.11 the scanty AEOD distribution at the surface of the 3D image show that the dust and haze contributes significantly in the aerosol content over the study area. In other words, the percentage of dust and haze can be roughly estimated using a well-planned computational technique. In this case, the author suggests 36%. The implication of this observation is that the anthropogenic pollution is the major contributor to aerosol distribution over West Africa. It can be observed that the background aerosol content (mainly dust and haze is high). However, some region (the sea at the coastal) demonstrated zero background aerosol content (Fig. 4.11). This means that the high haze level (Fig. 4.8) were not from ship activities at the coastal. Hence the haze was basically transport from the land. The 3D image also

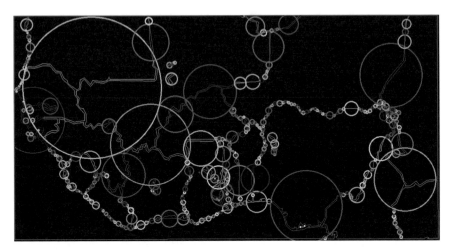

Fig. 4.10 Contour detection of satellite image

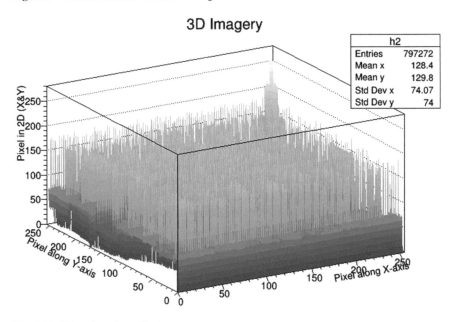

Fig. 4.11 3D setting of satellite image

shows a significant peak around Niger and Chad. This spot was detected in the image contour detection technique in Fig. 4.10.

Figure 4.12 has six main peaks and seven minor peaks. One very interesting concept of image projection is the region of peak (RoP) that depicts the line of highest impact of a given parameter along a specific axis. The main peaks are pixel-40 (Sierra Leone, Guinea, south-west Mali and Mauritania), 80 pixel (Cote d'Ivoire, Mali and Mauritania), pixel-120 (Ghana, Burkina Faso and Mali), pixel-140 (Benin

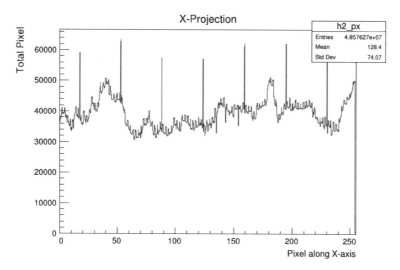

Fig. 4.12 X-projection of satellite image

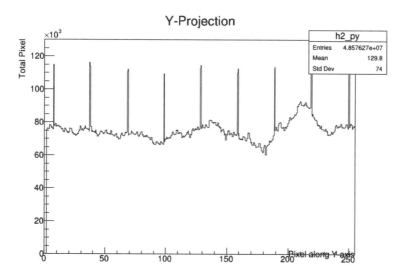

Fig. 4.13 Y-projection of satellite image

and Niger), pixel-180 (Nigeria and Niger) and pixel-250 (northern Cameroun and Chad). The minor peaks are the straight-tiny distinctive lines on Fig. 4.12. They are located on pixel-20 (Guinea, Gambia and Mauritania), pixel-55 (Liberia, Guinea, Mali and Mauritania), pixel-90 (Cote d'Ivoire, Burkina Faso and Mali), pixel-125 (boundary of Ghana and Togo, Burkina Faso and Mali), pixel-160 (Nigeria and Niger), pixel-195 (Nigeria and Niger) and pixel-230 (Cameroun, Nigeria and Chad) (Fig. 4.13).

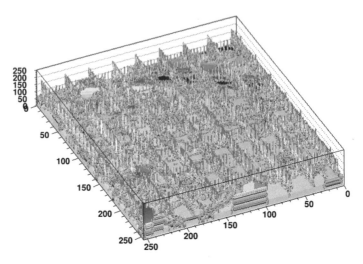

Fig. 4.14 The spectrum analysis of the satellite image

Unlike the X-projection, the Y-projection has one main peak and nine minor peaks. The main peak was at pixel-220 (Mauritania, Mali, Niger and Chad). The minor peaks are pixel-10 (sea), pixel-40 (coastal region of central Liberia, southern Cote d'Ivoire, southern Ghana, southern Togo, southern Benin, southern Nigeria and central Cameroun), pixel-60 (southern Sierra Leone, northern Liberia, central Cote d'Ivoire, central Ghana, central Togo, central Benin, central Nigeria and central Cameroun), pixel-102 (central Guinea, boundary of Mali and Cote d'Ivoire, southern of Burkina Faso, northern Ghana, northern Togo, northern Benin, northern Nigeria, northern Cameroun and central Chad), pixel-130 (Gambia, Guinea Bissau, Mali, Burkina Faso, south-east Niger, northern Nigeria and central Chad), 160 pixel (Senegal, Mali, northern Burkina Faso, Niger and Chad), pixel-190 (northern Senegal, Mauritania, Mali, Niger and Chad), pixel-220 (Mauritania, Mali, Niger and Chad), and pixel-240 (Mauritania, Mali, Niger and Chad).

Figure 4.14 describes how turbulent the mixing ratio of haze or dust or both in the region Fig. 4.8. It means that the particulates from haze and dust are uniform in the tropospheric profile of Fig. 4.8.

Aerosol optical depth is defined as a measure of the extinction of the solar beam by aerosol particulates from dust, haze, bush burning, gas flaring, industrial emission, building emission, automobile emission, domestic emission etc. Green band provides daytime observations related to the land, clouds and aerosols. It is used for air pollution studies and for estimating solar insolation. It is also used to estimate peak vegetation, which is salient for assessing plant vigor. The estimation of peak vegetation is possible when the green band is used to distinguish soil from vegetation. MIL3DAE (v 1, 2) is a product of the Multi-angle Imaging SpectroRadiometer (MISR) level 2 for land surface. It is prominent in determining the component global

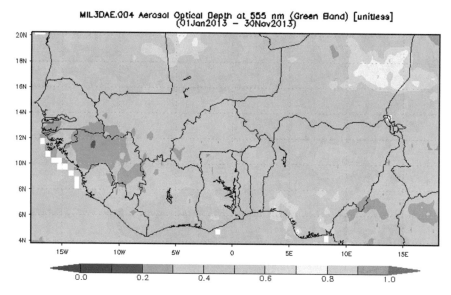

Fig. 4.15 Satellite image AOD at 555 nm, Jan. to Nov., 2013

aerosol product covering per day. Hence, the information on Fig. 4.15 illustrates AOD during the day over the study area.

In Fig. 4.15, the influence of Sahara dust during the day over Mauritania Mali, Niger and Chad. Niger has the highest concentration of dust during the day. Also, it can be inferred from Fig. 4.15 that the concentration of anthropogenic pollution source is from the southern Nigeria, coastal part of Ghana and Cote d'Ivoire. Also, the region of very low AOD can be seen in large area of Guinea and minute parts of Nigeria, Cameroun, Senegal, Mali, Burkina Faso and Cote d'Ivoire.

Figure 4.16 show a clearer color representation of the satellite image. It is very easy to understand the sources of day pollution and the dispersion path that can be described by the red circle. The purple line describes the second concentric aerosol flow path. The speed of the aerosol layers from the source to the secondary destination has been calculated by Emetere (2017). The sources of the daylight pollution had been discussed in Fig. 4.15. However, the dispersion path of aerosol during the day shows that it cuts-through Nigeria. Togo, Benin, Ghana, Cote d'Ivoire, Liberia into the coastal path of Sierra Leone, Guinea, Guinea Bissau, Gambia, Senegal, and Mauritania. The path continues inland Mauritania, Mali and Niger.

Figure 4.17 show the different contour within each big circles. It is observed that the circles describe the aerosol flow path that was discussed in Fig. 4.16. Figure 4.18 show that some geographical areas does not have background aerosol concentration (BAC). The BAC for the affected location is very high. Hence, like the AEOD, the BAC is high at the lower atmosphere of Fig. 4.15.

The X-projection has ten peaks at pixel-40 (Sierra Leone, Guinea, Mali and Mauritania), pixel-80 (Cote d'Ivoire and Mali), pixel-95 (Cote d'Ivoire, Burkina Faso and

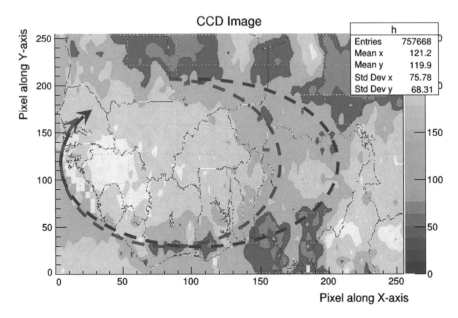

Fig. 4.16   x and y pixel redefinition of the satellite image

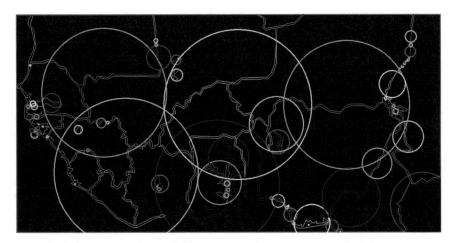

Fig. 4.17   Contour detection of satellite image

Mali), pixel-105 (Cote d'Ivoire, Burkina Faso and Mali), pixel-125 (Ghana, Burkina Faso and Mali), Pixel-135 (Togo, Burkina Faso and Mali), pixel-175 (Nigeria and Niger), pixel-185 (Nigeria and Niger), pixel-200 (Cameroun, Nigeria and Niger) and pixel-220 (Cameroun, Nigeria and Niger). The highest peak can be found at peak-40. Mali, Niger and Burkina Faso were the most affected cities based on the frequency of the location mentioned under the peak. It was observed that all the nations along the aerosol flow path (Fig. 4.16) were mentioned on all the peak location. Hence, it

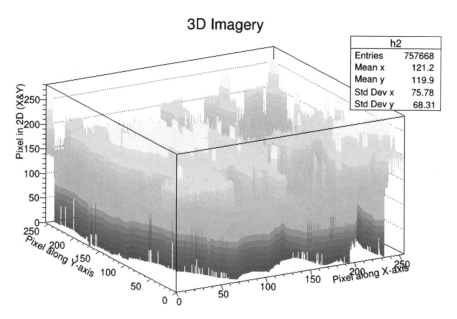

Fig. 4.18 3D setting of satellite image

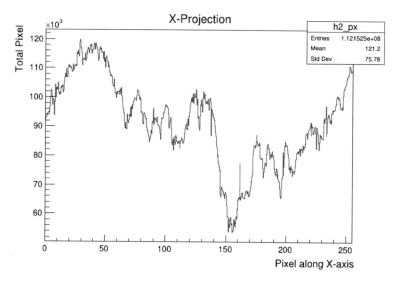

Fig. 4.19 X-projection of satellite image

is trivial to infer that the peak location may also describe the aerosols flow path. The sinusoidal nature of the image feature in Fig. 4.19 may be due to the very unique aerosol flow path for AOD 550 nm.

Figure 4.20 describes the Y-projection. It has three major peaks and two minor peaks. The major peaks are pixel-5 (sea), pixel-45 (Liberia, southern Cote d'Ivoire,

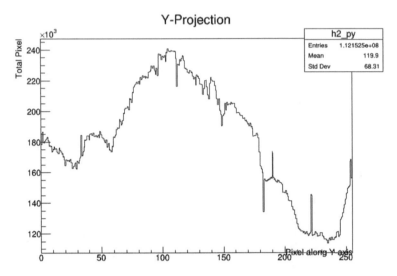

Fig. 4.20   Y-projection of satellite image

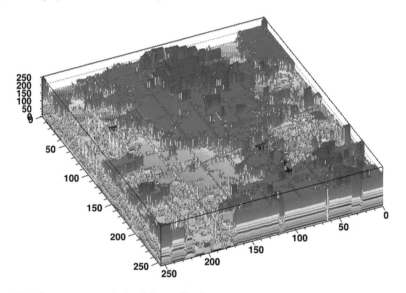

Fig. 4.21   The spectrum analysis of the satellite image

southern Ghana, southern Togo, southern Benin, southern Nigeria and central Cameroun) and pixel-110 (Guinea, southern Mali, southern Burkina Faso, northern Ghana,
northern Togo, northern Benin, northern Nigeria, northern Cameroun and southern
Chad). The minor peaks are pixel-40 (Senegal, southern Mali, northern Burkina Faso,
northern Nigeria and Chad) and pixel-220 (Mauritania, Mali, Niger and Chad). From
the intersection rule, pixel-220 both on X and Y projections show that the Niger is
the most affected (Fig. 4.21).

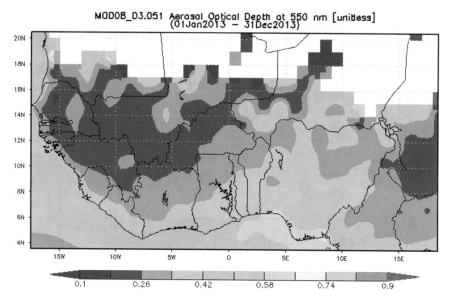

Fig. 4.22 Satellite image AOD at 550 nm, Jan. to Dec., 2013

Figure 4.12 show that the vertical profile of AOD during the day is fairly stable in some parts. This could be explained by the kind of wind dynamics that has been reported over West Africa. The turbulence in the vertical profile can be seen in Niger, northern parts of Nigeria, Chad, northern Cameroun, Mauritania, Senegal and Liberia.

AOD 550 was obtained from the Moderate Resolution Imaging Spectroradiometer (MODIS) Level 2 (MOD08_D3.051). In Fig. 4.22, it was observed that the data in the upper part was distorted. Hence, the information on the Mauritania, Mali, Niger and Chad was not given. The MODIS satellite image show large aerosols concentration at the coastal parts of Nigeria and over the sea. Figure 4.23 show vividly the high AOD concentration observed at the sites of Sierra Leone and Liberia. Secondly, the aerosol flow parts can be seen (red circle) which corroborates the observations in Fig. 4.16. The second aerosols flow path can be seen (purple line). Like Fig. 4.16, Guinea, northern Sierra Leone, northern Liberia, northern Cote d'Ivoire, northern Ghana, Burkina Faso and southern Mali has low AOD during the day (Fig. 4.23). Despite the distortion in Fig. 4.22, it was observed that some parts in the re-processed image (Fig. 4.23) still show high concentration of aerosols in Mauritania, Mali and Niger.

Figure 4.24 shows three large circles over Nigeria, Benin, Togo, Ghana and eastern Cote d'Ivoire. This means that aside the information that is discussed in Fig. 4.23, there are other hotspots that may not be adequately illustrated. This observation is very important in planning ground measurement over a geographical area that the researcher do not have pre-information on Emetere (2016b). Figures 4.18 and 4.25

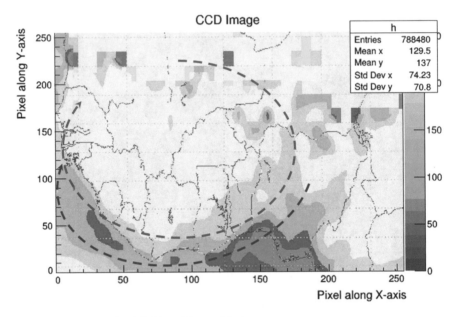

Fig. 4.23 x and y Pixel redefinition of the satellite image

Fig. 4.24 Contour detection of satellite image

shows that the sea does not have background aerosol concentration (BAC). The BAC for the inland is equally high as shown in Fig. 4.25. Contrary to Fig. 4.18, the higher troposphere is characterized by low AOD.

Figure 4.26 has nine peaks i.e. pixel-22 (Guinea Bissau, Gambia, Senegal and Mauritania), pixel-40 (Sierra Leone, Guinea, southern Mali and Mauritania), pixel-65 (Liberia, Guinea, Mali and Mauritania), pixel-80 (Cote d'Ivoire and Mali), pixel-

3D Imagery

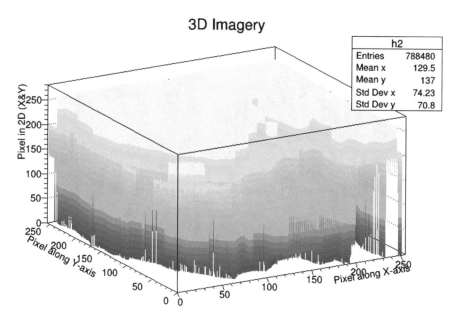

h2	
Entries	788480
Mean x	129.5
Mean y	137
Std Dev x	74.23
Std Dev y	70.8

Fig. 4.25 3D setting of satellite image

110 (Ghana, Burkina Faso and Mali), pixel-120 (Ghana, Burkina Faso and Mali), pixel-145 (Nigeria, northern Benin, south Niger and Mali), pixel-165 (Nigeria and Niger) and pixel-215 (Cameroun, Nigeria and Niger). From the frequency of location observed on each peak, it could be concluded that the highest concentration can be found in Mali, Niger and Nigeria.

In Fig. 4.27, two peaks were observed. The peaks were pixel-120 (Guinea Bissau, Guinea, southern Mali, Burkina Faso, northern Benin, Nigeria, northern Cameroun and Chad), pixel-200 (Mauritania, Mali, Niger and Chad). The intersection rule shows that the location with the highest AOD at 550 nm were Burkina Faso and Mali. The inclusion of distortion in the satellite image may compromise the image re-processing technique. For example, the inclusion of Burkina Faso as region of highest AOD is false because of the preceding results shown in Figs. 4.22 and 4.23.

Figure 4.28 shows the turbulence in the vertical profile of Fig. 4.22. It should be noted that the distortion of the AOD image over Mauritania, Mali, Niger and Chad could compromise the turbulence zones in Fig. 4.28. Through the AOD source over Nigeria, vertical turbulence could be seen in the coastline of Nigeria, Benin, Ghana, Cote d'Ivoire and Liberia. It was also observed that the turbulence extended inland towards Guinea. This further confirms that aerosol transport may follow the Gaussian path as expressed in the Gaussian plume model (Walcek 2004; Tirabassi et al. 2010).

The aerosol absorption optical depth (AAOD) at 500 nm was obtained from OMI OMAERUVd v003. OMI OMAERUVd v003 are popularly known as the OMI-Aura level-3. It is an enhanced algorithm developed by NASA Goddard Earth Sciences Data and Information Services Center (GES DISC). AAOD are used to calculate the

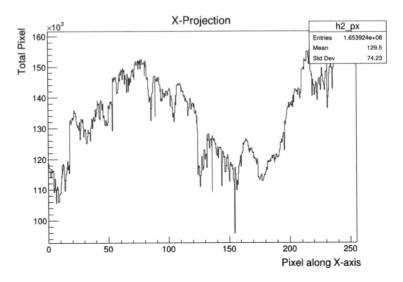

Fig. 4.26 X-projection of satellite image

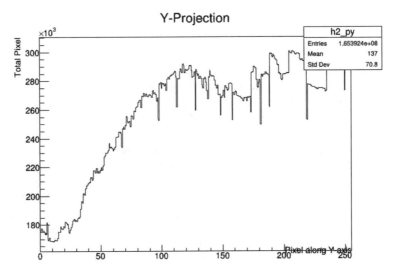

Fig. 4.27 Y-projection of satellite image

radiative forcing of black carbon (BC) aerosols (Bond et al. 2013; Koike et al. 2014). Bond and Bergstrom (2006) defined absorbing aerosol as soot. In this section, the type of AOD considered are the soot or black carbon obtained from bush burning, gas flaring, industrial burning etc. Figure 4.29 shows that there is a large concentration of black carbon in the southern parts of Nigeria. Some parts of Benin and Cote d'Ivoire were observed to have high concentration of black carbon. It was observed that the sea close to Cape Verde and Chad have high concentration of black carbon. Also,

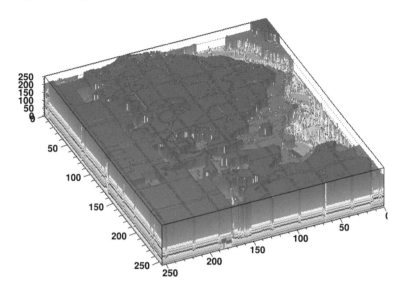

Fig. 4.28 The spectrum analysis of the satellite image

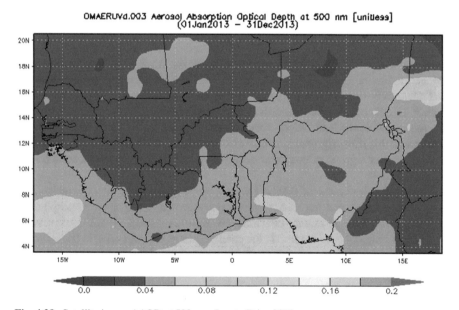

Fig. 4.29 Satellite image AAOD at 500 nm, Jan. to Dec., 2013

about 57% of the study area have low concentration of black carbon. The re-processed image is shown in Fig. 4.30.

In Fig. 4.30, it was observed that the processed image merged two colours located at concentration of 0.04. This makes it very clear that about 92% of the study area

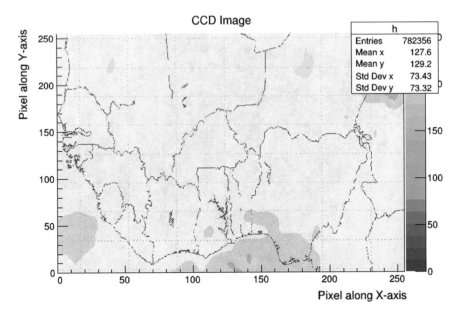

Fig. 4.30   x and y pixel redefinition of the satellite image

have low concentration of black carbon. Hence, the black carbon contributes the lowest fraction of aerosols over West Africa. Figure 4.31 show very striking features that may not be described in the satellite image. First, the circle located at the north-central of Nigeria. A visible contour can be observed inside the circle. Secondly, the red circle over Chad, Nigeria and Niger. Thirdly, two circles over Mali and Mauritania that show significant contours. Other circles were not recon with because it is either over a group of countries or tiny spots along boundaries between two or more countries.

Figure 4.32 shows that the BAC of the lower atmosphere in the region is reasonably high. It also shows that the black carbon concentration in the upper troposphere is very low. Also, it shows that the coastline towards the sea had no BAC. Like previous results, the imagery over the sea on the satellite image is largely due to the influx of aerosol from a primary source(s) close to the coastline.

Figure 4.33 show no significant peak. This means that the black soot concentration is cumulatively low and uniform along the x-axis of the satellite image. However, the Y-projection shows a large variation (Fig. 4.33) with eleven peaks. The peaks are pixel-10 (sea), pixel-20 (coastline), pixel-40 (southern Liberia, southern Cote d'Ivoire, southern Ghana, southern Togo, southern Benin, southern Nigeria and northern Cameroun), pixel-55 (southern Sierra Leone, northern Liberia, southern Guinea, central Cote d'Ivoire, central Ghana, central Togo, central Benin, central Nigeria, northern Cameroun and southern Chad), pixel-70 (Sierra Leone, Guinea, Cote d'Ivoire, Ghana, Togo, Benin, Nigeria, Cameroun and Chad), pixel-85 (Sierra Leone, Guinea, Cote d'Ivoire, Ghana, Togo, Benin, Nigeria, Cameroun and Chad),

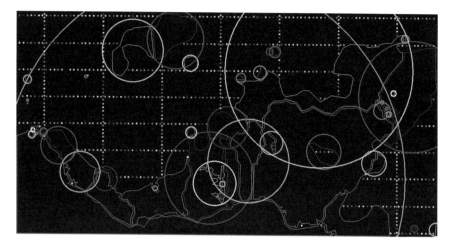

Fig. 4.31 Contour detection of satellite image

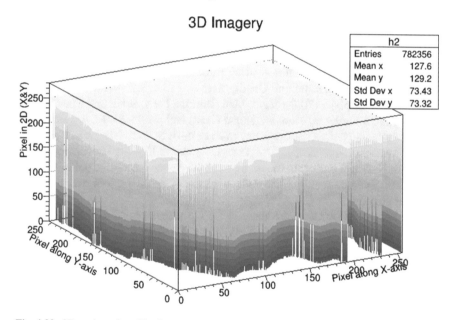

Fig. 4.32 3D setting of satellite image

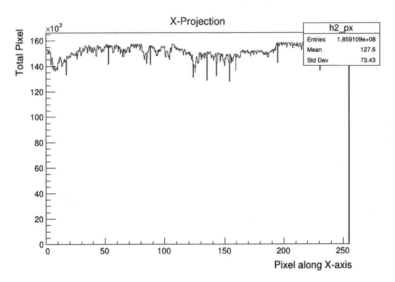

Fig. 4.33 X-projection of satellite image

pixel-120 (Guinea Bissau, Guinea, Mali, Burkina Faso, northern Benin, northern Nigeria, northern Cameroun and Chad), pixel-140 (Senegal, Mali, Burkina Faso, Niger and Chad), pixel-170 (Senegal, Mali, Burkina Faso, Niger and Chad), pixel-185 (Senegal, Mauritania, Mali, Niger and Chad) and pixel-220 (Mauritania, Mali, Niger and Chad). Neither the frequency rule nor the intersection rule can be applied in this sense since one of the 2D axis (x-axis) shows that the distribution of the black soot is almost uniform. The vertical profile can also be adjudged as uniform (Fig. 4.35). It can be seen that the vertical-profile of the black soot over the study fluctuates slightly.

The AAOD at 388 nm was considered in Fig. 4.36. Recall that the AAOD at 500 nm was earlier discussed in Figs. 4.29, 4.30, 4.31, 4.32, 4.33, 4.34 and 4.35. 388 nm means that the satellite image is been viewed from a closer perspective. Hence, this Figure is insightful to validate the vertical profile analysis of previous figures. First, the satellite image (Fig. 4.36) shows that below the vertical profile of Fig. 4.29, high source of black carbon can be seen between the boundaries of Mauritania and Mali. Secondly, more area under the image shows higher black carbon concentration. This shows that the area has significant presence of black carbon in its background. The transport or dispersion of the black carbon from its source to the upper troposphere or lower stratosphere depends on the wind dynamics, concentration level of black carbon, life-time and existing aerosol retention in the upper troposphere.

Figure 4.37 shows a clearer picture of aerosol flow path that was discussed earlier. The new color representation helps to identify location of unique features. For example, it can be clearly seen that there are six spots (first spot at the offshore of Ghana, Togo and Benin, second spot at the offshore of Nigeria, third spot at the boundary of Mauritania and Mali, fourth spot in Mali, fifth spot in Niger, and sixth spot at the

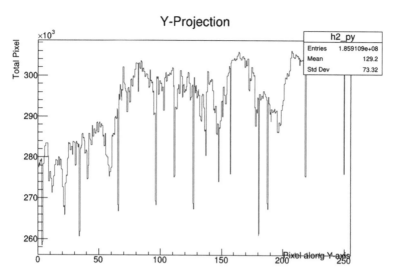

Fig. 4.34 Y-projection of satellite image

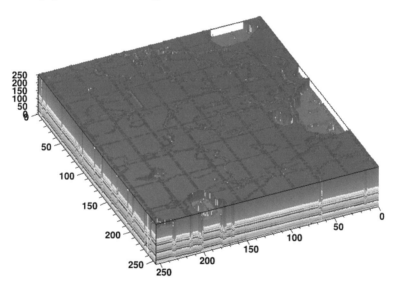

Fig. 4.35 The spectrum analysis of the satellite image

boundaries of Niger and Chad) that may show the source of black carbon. This image shows more interesting black carbon deposition in the coastline of West Africa.

Figure 4.38 shows the five drawing defect as shown in Fig. 4.36. The sensitivity of the contour detection is high and can be used to understand the bigger circles within the study area. For example, the green (biggest) circle shows a unique trend of black carbon catchment in Fig. 4.36. Comparing Figs. 4.38 and 4.31, it can be observed

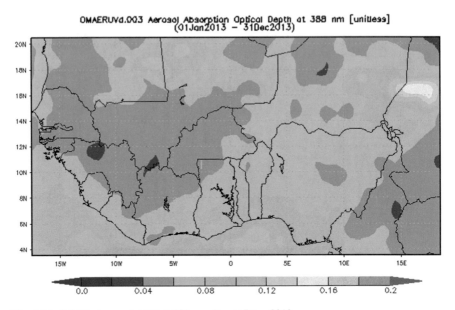

Fig. 4.36 Satellite image AAOD at 388 nm, Jan. to Dec., 2013

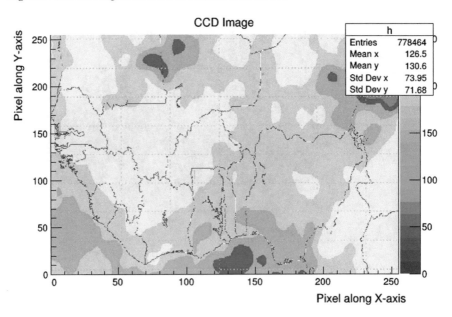

Fig. 4.37 x and y pixel redefinition of the satellite image

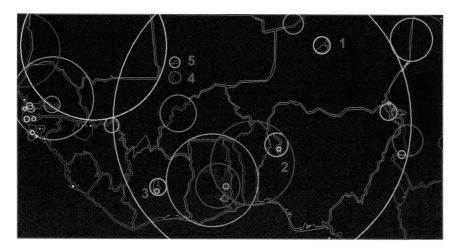

Fig. 4.38 Contour detection of satellite image

that the contour detection pattern differs. This means that even at different vertical profile, layers of aerosols (which are invisible to the naked eye) can be identified.

The BAC in Figs. 4.32 and 4.39 are almost similar. This affirms that the 3D image could give the approximate BAC of an image. The scanty top of Fig. 4.39 describes that there is more concentration of black carbon at the mid troposphere. The X-projection of Fig. 4.40 show more activity compared to Fig. 4.33. It has a sinusoidal feature that depicts positional variation along the x-axis.

Figure 4.40 has ten peaks i.e. pixel-35 (Sierra Leone, Guinea, Senegal and Mauritania), pixel-60 (Liberia, Guinea, southern Mali and Mauritania), pixel 70 (Cote d'Ivoire, Mauritania and Mali), pixel-90 (Cote d'Ivoire, Burkina Faso and Mali), pixel-100 (Cote d'Ivoire, Burkina Faso and Mali), pixel-110 (Ghana, Burkina Faso and Mali), pixel-130 (Togo, Burkina Faso, southern Niger and Mali), pixel-170 (Nigeria and Niger), pixel-185 (Nigeria and Niger), pixel-200 (Cameroun, Nigeria and Niger). From the frequency rule, Mali had the highest concentration of black carbon on the x-projection. Figure 4.41 show the Y-projection of the satellite image.

There are five peaks in Fig. 4.41. They are pixel-20 (coastline of Liberia, coastline of Cote d'Ivoire, coastline of Nigeria and Cameroun), pixel-90 (Sierra Leone, Guinea, Cote d'Ivoire, Ghana, Togo, Benin, Nigeria, and Cameroun), pixel-100 (Guinea, Cote d'Ivoire, Ghana, Togo, Benin, Nigeria, northern Cameroun and Chad), pixel-140 (Senegal, Mali, Burkina Faso, southern Niger, Nigeria and Chad) and pixel 165 (Senegal, Mali, Burkina Faso, Niger and Chad). By the frequency rule, Nigeria has the highest concentration along the y-axis. By the intersection rule, Pixel-90 on both (x and y) projections (Cote d'Ivoire) and pixel-100 on both projections (Cote d'Ivoire). The spectrum analysis of the satellite images (Fig. 4.42) show the high perturbation in the vertical profile. More so, the aerosol flow path had higher perturbation in the vertical profile.

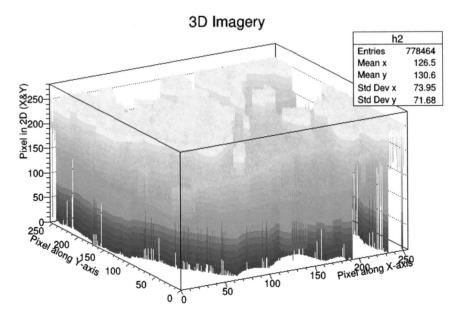

Fig. 4.39 3D setting of satellite image

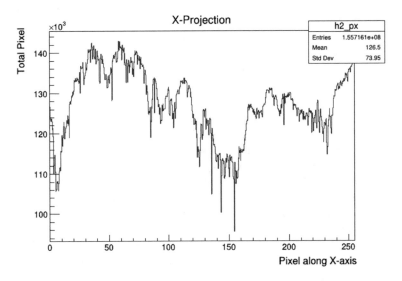

Fig. 4.40 X-projection of satellite image

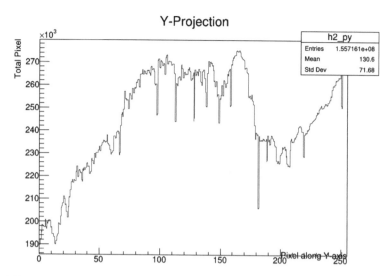

Fig. 4.41 Y-projection of satellite image

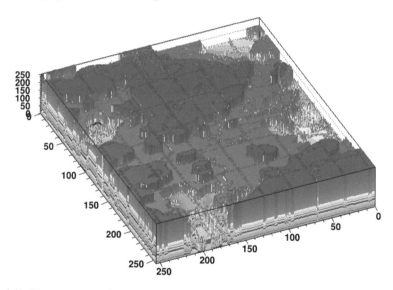

Fig. 4.42 The spectrum analysis of the satellite image

AEOD at 388 nm had been discussed in Fig. 4.8. Figure 4.43 show the AEOD at 500 nm. Generally, AEOD describes the level of impact on direct and indirect aerosol forcing (Lacis and Mishchenko 1995). A type class of aerosol that affects AEOD is the tropospheric sulfate (Tegen et al. 1997). Figure 4.43 show that the region of highest concentration was Chad, Niger, Mali and Mauritania. Coincidentally, the same region had high black carbon. Hence the anthropogenic pollution of the location will be very interesting to study. Figure 4.44 clearly show the aerosol flow path of the

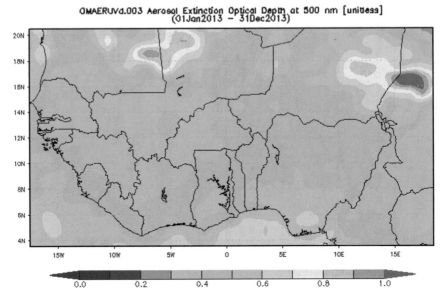

Fig. 4.43 Satellite image AEOD at 500 nm, Jan. to Dec., 2013

satellite image. Like the AAOD—Fig. 4.37, the sea at the coastline of West Africa had high AEOD. Hence Fig. 4.44 shows seven hotspots i.e. south-south of Nigeria (hotspot 1), offshore of Nigeria, Benin, Togo and Ghana (hotspot 2), offshore of Cote d'Ivoire (Hotspot 3), offshore of Sierra Leone and Liberia (Hotspot 4), boundary of Niger and Chad (hotspot 5), boundary of Niger and Mali (hotspot 6) and boundary of Mali and Mauritania (hotspot 7). The contour detection in Fig. 4.45 show same big circle (like Fig. 4.38) that describe the aerosol flow path.

Figure 4.46 shows a high BAC inland and no or little BAC offshore. From the surface of the 3D image, it can be inferred that the extinction aerosol is generally high. The X-projection has six peaks i.e. pixel-35 (Sierra Leone, Guinea, Senegal and Mauritania), pixel-75 (Cote d'Ivoire, Mali and Mauritania), pixel-110 (Ghana, Burkina Faso and Mali), pixel-130 (Togo, Burkina Faso, southern Nigeria and Mali), pixel-180 (Nigeria and Niger) and pixel-205 (Cameroun, Nigeria and Niger). By the frequency rule, Niger has the highest AEOD. Figure 4.48 show that Y-projection has a Gaussian distribution with its peak at pixel-120 (Guinea Bissau, Guinea, Mali, Burkina Faso, northern Benin, Nigeria, northern Cameroun and Chad) (Fig. 4.47).

However, there were some minor peaks along the Gaussian distribution located at pixel-35 (Liberia, Cote d'Ivoire, Togo, Benin, Nigeria and Cameroun), pixel-185 (Senegal, Mauritania, Niger, Mali and Chad) and pixel-220 (Mauritania, Mali, Niger and Chad). The frequency and the intersection rule cannot hold because of the Gaussian distribution. Figure 4.49 show the degree of vertical profile turbulence. Therefore, AEOD at 500 nm show how much the area is influenced by sulfate aerosols.

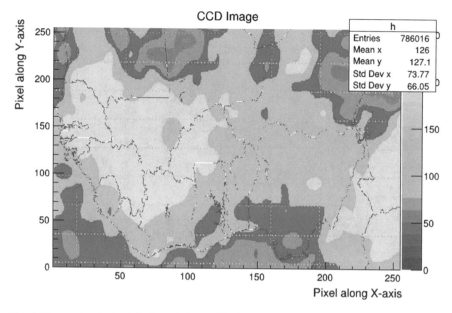

Fig. 4.44 x and y pixel redefinition of the satellite image

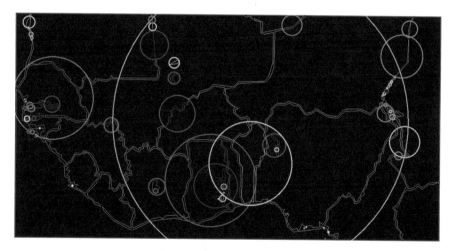

Fig. 4.45 Contour detection of satellite image

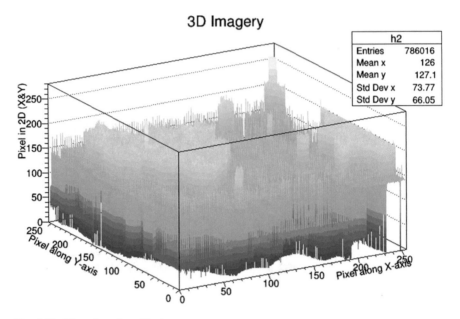

Fig. 4.46 3D setting of satellite image

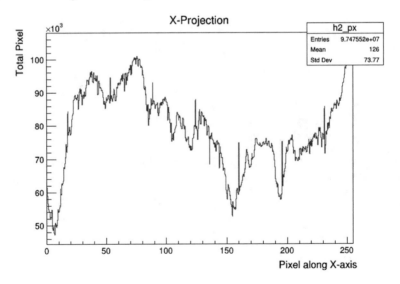

Fig. 4.47 X-projection of satellite image

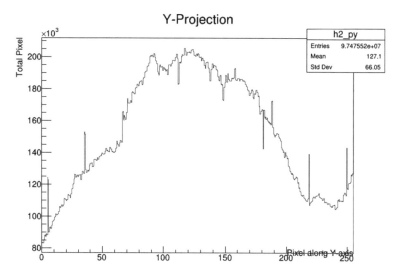

Fig. 4.48 Y-projection of satellite image

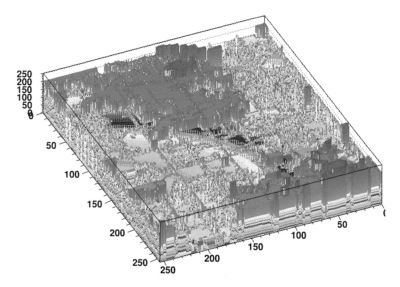

Fig. 4.49 The spectrum analysis of the satellite image

4.2 Implication of Computational Outcomes

In the last section, it can be inferred that the image re-processing technique draws-out more information about the satellite image in that: it re-maps the original satellite image (in distinctive colours) for better understanding of its hidden features; it shows aerosol concentration hotspots (via contour detection) that may not be seen by the ordinary eye; it provides the background aerosol concentration (BAC) of an area and spatial distribution of aerosols at the upper troposphere/lower stratosphere; it shows the image projections in the x and y axis; and it shows the degree of vertical profile turbulence.

It was also discussed in the last section that the most active regions of a satellite image can be determined by the frequency and intersection rule of the locations listed under the peaks of the X and Y projection. The peaks in the X and Y projection were classified as major or minor peaks. For example, in Figs. 4.12 and 4.13, the worst hit places by dust and haze are Mauritania, Mali, Niger and Chad; followed by Burkina Faso, Cote d'Ivoire, Ghana, Togo, Benin, Nigeria and Cameroun. By the intersection rule (i.e. major or minor peaks at ≤ 5), the close peak intersection in both the X and Y projections are pixel-55 (X-projection) and pixel-60 (Y-projection), pixel-125 (X-projection) and pixel-130 (Y-projection), pixel-160 (X-projection) and pixel-160 (Y-projection), pixel-195 (X-projection) and pixel-190 (Y-projection). Hence, from the intersection technique, the country that appear in both intersection points are Liberia (first intersection, Burkina Faso and Mali (second intersection), Niger (third intersection), and Niger (fourth intersection). Therefore, it is very logical to conclude that the country that has the highest concentration of dust and haze is Niger. However, going back on Fig. 4.8, it looks contradictory that the AEOD over Guinea was among the highest but the AOD at 550 nm was the lowest. This contradiction may be explained with respect to the unique transports of aerosol particulates over West Africa (Emetere 2017). Hence, aerosol layers travel from one neighbouring country or community to another.

One of the salient lessons learnt by re-processing satellite images is that aerosol move in a concentric path (red circle in Fig. 4.16) that was christened 'aerosol flow path'. From Figs. 4.16, 4.23, 4.37 and 4.44, the route of the aerosol flow path is from Nigeria, Benin, Togo, Ghana and Cote d'Ivoire—towards the coastal parts of the south-western countries on Fig. 4.15. The route of the aerosol flow path from the north is through Mauritania, Mali, Niger and Chad into northern Nigeria. Two aerosol flow paths were notice over Nigeria. The first aerosol flow path is through the north-west, north central and south west of Nigeria. The second aerosol flow path is through the north-east, north-central, south-east and south-south of Nigeria. This informs our consideration of the data-processing locations in the next chapter.

Figures 4.15 (AOD at 555 nm) and 4.22 (AOD at 550 nm) affirm that the hotspots were located in Ghana and Cote d'Ivoire. Secondly, the AOD (at 555 nm) show that aerosol flow path route along the coastal part of Nigeria and extends inland (Figs. 4.15 and 4.16). Thirdly, AOD 550 nm show a significant extension of the AOD over the sea. In Fig. 4.23, the aerosol flow path can be seen (red circle) which corroborates the observations in Fig. 4.16. The turbulent locations described in Fig. 4.21 do not correspond to the locations mentioned under the aerosol flow path in Fig. 4.16. This suggests that the aerosol flow path takes place at the upper troposphere while the turbulence takes place at the lower atmosphere. Hence, the image re-processing could afford salient information that may not be described on satellite images.

The implication of regions without BAC (over sea) as presented in Figs. 4.18 and 4.25 show that aerosol pollution that is portrayed in the satellite imagery may not be due to movement of ship across the coastal part, but the aerosols transport from pollution sources close to the coastline. The BAC in Fig. 4.32 and 4.39 are almost similar. This affirms that the 3D image could give the approximate BAC of an image.

The main advantage of the re-processed image is its ability to encourage comparative study between two location or same location at different times. In the light of the above, the satellite imagery of the following was processed:

   i. Satellite image AAOD at 500 nm, Jan. to Dec., 2007 (Appendix: Figs. A.1–A.7);

  ii. Satellite image AEOD at 388 nm, Jan. to Dec., 2007 (Appendix: Figs. A.8–A.14);

 iii. Satellite image AOD at 555 nm, Jan. to Dec., 2007 (Appendix: Figs. A.15–A.21);

 iv. Satellite image AOD Pixel Counts, Jan. to Dec., 2007 (Appendix: Fig. A.12–A.28);

  v. Satellite image AOD, Coarse mode, Jan. to Dec., 2007 (Appendix: Figs. A.29–A.35);

 vi. Satellite image AAOD at 388 nm, Jan. to Dec., 2007 (Appendix: Figs. A.36–A.42);

 vii. Satellite image ACOD at 550 nm, Jan. to Dec., 2007 (Appendix: Figs. A.43–A.49);

viii. Satellite image ACOD at 550 nm-FM, Jan. to Dec., 2007 (Appendix: Figs. A.50–A.56);

 ix. Satellite image AEOD at 500 nm, Jan. to Dec., 2007 (Appendix: Figs. A.57–A.63);

  x. Satellite image AOD at 550 nm, Jan. to Dec., 2007 (Appendix: Figs. A.64–A.70).

In the comparative analysis, the satellite image of dust aerosol column optical depth (ACOD) at 550 nm, for the year 2007 (Appendix: Fig. A.43) was used as the standard because dust is the main aerosol content over the research area (Senghor et al. 2017). Figure 4.50 show the X-projection of two types of aerosol parameter [i.e. AAOD at 500 nm for 2007 (indigo) and AAOD at 500 nm for 2013 (blue)] and the standard [i.e. ACOD at 550 nm for 2007 (yellow)]. It can be inferred from Fig. 4.50 that there is about 14.3% increase in AAOD from 2007 to 2013. Second, the AAOD over locations in pixel-205 (north-west Cameroun, north-east Nigeria and Niger) was constant between 2007 and 2013. Third, ACOD cuts across: AAOD-2013 at pixel-18 (Guinea Bissau, Gambia, Senegal and Mauritania); AAOD-2007 at pixel-50 (Liberia, Guinea, Mali and Mauritania); AAOD-2007 at pixel-135 (Benin, Niger and Mali); pixel-145 (Nigeria, Benin, Niger and Mali); pixel-155 (Nigeria and Niger). The intersection points are certainly where the ACOD equals the AAOD. Hence, it could be inferred that AAOD are prominent over the region.

In Appendix-Fig. A.71, the AEOD at 388 nm were almost similar. The ACOD interacted with AEOD (2007 and 2013) between longitude 17°W and 9°W. In Appendix-Fig. A.72, the AOD at 555 nm was nearly similar. The ACOD interacted with AOD (2007 and 2013) throughout the X-projection. AOD pixel count for 2007 and 2013 was similar (Appendix-Fig. A.73). The ACOD interacted with AEOD

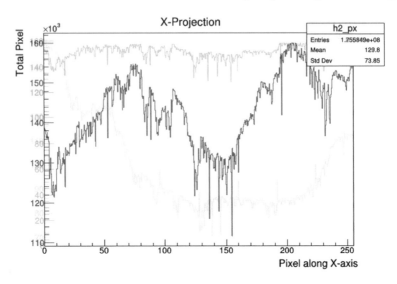

Fig. 4.50 Comparative analysis of AAOD at 500 nm (2007 and 2013) and assumed standard (ACOD 2007) in the X-projection

(2007 and 2013) between longitude 17°W and 14°W. There was less than 2% increase between the AAOD (at 388 nm) in the year 2007 and 2013 (Appendix-Fig. A.74). The ACOD interacted with AAOD at 388 nm (2007 and 2013) between longitude 17°W and 13°W. In Appendix-Fig. A.75, the difference between the AEOD at 500 nm for the year 2007 and 2013 was ≤35%. The ACOD interacted with AEOD (2007 and 2013) throughout the X-projection. Almost same analysis was for Appendix-Fig. A.76 as Appendix-Fig. A.75.

The AAOD increase for each country can be estimated by multiplying the percentage increase/decrease between the two years (2007 and 2013) and AAOD values. For example in pixel-60 for AAOD in 2007, the countries and the corresponding AAOD is Liberia (≈0.08), Guinea (≈0.06), Mali (≈0.05) and Mauritania (≈0.04). The percentage increase at pixel-60 is 6.7%. The multiplication factor is (100 + 6.7)/100. Hence the new value expected in 2013 will be Liberia (≈1.067 * 0.08 = 0.085 along the x-axis), Guinea (≈1.067 * 0.06 = 0.064 along the x-axis), Mali (≈1.067 * 0.05 = 0.053 along the x-axis) and Mauritania (≈1.067 * 0.04 = 0.043 along the x-axis).

The line and color arrangement in Fig. 4.51 is the same as Fig. 4.50. The standard (ACOD 2007) describes a phase-shift parabola while both AAOD (2007 and 2013) exponential-growth parabola. There was no major difference between AAOD for 2007 and 2013. The meeting point of the AAOD and ACOD is at pixel-75 (Cote d'Ivoire, Mali and Mauritania). The Y-projection comparison is shown for AEOD at 388 nm (2007 and 2013), AOD at 555 nm (2007 and 2013), AOD pixel count (2007 and 2013), AAOD (2007 and 2013), AAOD at 388 nm (2007 and 2013), AAOD at 500 nm (2007 and 2013), and AEOD at 500 nm (2007 and 2013). The images are located at Appendix: Figs. A.77–A.82. The AEOD at 388 nm was almost similar with

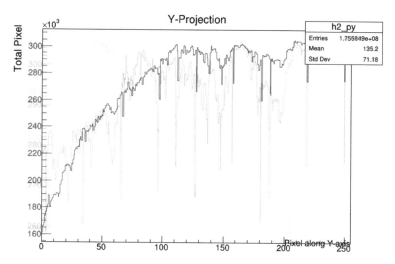

Fig. 4.51 Comparative analysis of AAOD at 500 nm (2007 and 2013) and assumed standard (ACOD 2007) in the Y-projection

AEOD 500 nm. However, there was a large difference between 2007 and 2013 of AEOD at 388 nm (Appendix: Fig. A.77). The percentage increase from 2007 to 2013 was ≤26%. In Appendix: Fig. A.78, both years (2007 and 2013) had Gaussian-like shape. The percentage increase from 2007 to 2013 was ≤33%.

Lastly in this section, the difference between images using the image intensity, mean pixel, standard deviation and full width at half maximum (FWHM) is illustrated below. The satellite image numbers are denoted as follows:

0—AOD pixel count at 388 nm (2013);
1—AEOD at 388 nm (2013);
2—AOD at 555 nm (2013);
3—AOD at 550 nm (2013);
4—AAOD at 500 nm (2013);
5—AAOD at 388 nm (2013);
6—AEOD at 500 nm (2013);
7—AAOD at 500 nm (2007);
8—AEOD at 388 nm (2007);
9—AOD at 555 nm (2007);
10—AOD Pixel Counts (2007);
11—AOD, Coarse mode (2007);
12—AAOD at 388 nm (2007);
13—ACOD at 550 nm (2007);
14—ACOD at 550nm-FM (2007);
15—AEOD at 500nm (2007);
16—AOD at 550nm (2007).

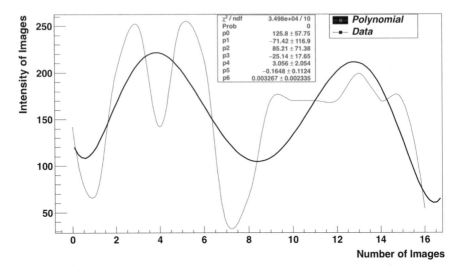

Fig. 4.52 Intensity of satellite images

Figure 4.52 is the intensities of the highlighted images above. It is observed that AOD at 550 nm (2013) and AAOD at 388 nm (2013) had the highest intensities. This means that the above mentioned images have the most significant value of a pixel in an image matrix. The polynomial values are listed to aid optimization when the environmental images are sequential in only one type of dataset.

Figure 4.53 is the mean pixel of the satellite images. AAOD at 388 nm (2007) had the highest image mean-pixel. Hence, the image has the most significant smallest addressable element of the image. AEOD at 500 nm (2013) and AEOD at 500 nm (2007) also had a high mean pixel. In Fig. 4.54, AOD pixel count at 388 nm (2013) had the highest standard deviation among images. Also in Fig. 4.55, AOD pixel count at 388 nm (2013) had the highest standard deviation among images.

4.3 Planning the Code Design

In this section, the discussion shall be on how to design the code used for running the above simulations. Planning the design of a code is paramount to the desired goal of the research (i.e. using computational process). Codes may be written in macro or compiled program. A program is an executable script that assists the computer system to execute specific instructions. The executable instructions are compiled using computer language such as C, C++ etc. The executable instructions are arranged in a modular manner, with good separation of subsections. Macro is usually something more simple and ad hoc. The common usage of macros is in software applications for mapping user input to different set of user output. Program makes use of larger tools

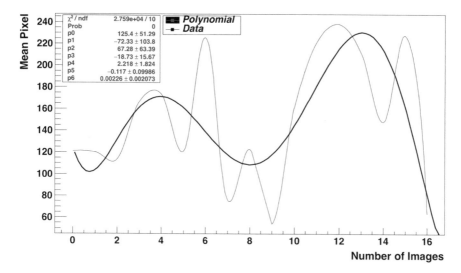

Fig. 4.53 Mean pixel of satellite images

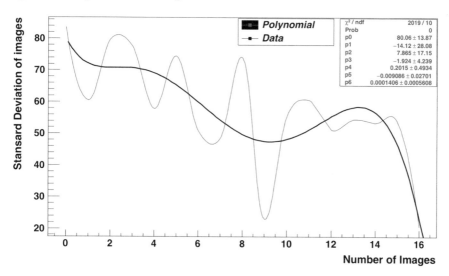

Fig. 4.54 Standard deviation of satellite images

e.g. Makefile, executable file, programme which are defined in '.cpp', '.o' and '.d'
while macro makes use of smaller tools which are more easily developed as scripts.

When planning to write a program, the following rules are essential:

  i. understand your application programme very well;
 ii. compile or build compatible libraries for your job specification;
iii. know how to link your libraries and application. The adopted Makefile are key
      towards a successful link;

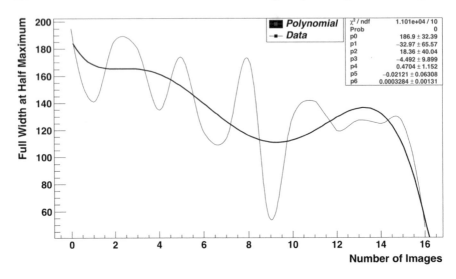

Fig. 4.55 Full width at half maximum of satellite images

iv.   Choice of compiler for your platform is very essential.

For example, the program used to illustrate analysis in this book was CERN-Root and the libraries were Tiff and OpenCV. The modalities for arranging the programme are:

  i. Write an appropriate header files
 ii. Write the application and library linker
iii. State the global parameters
 iv. State the prototypes
  v. Decide on the type of main program you want. Most programmer use the argc-argv functionality
 vi. Define your input, set-up and root files
vii. Choose carefully the process of uploading the input image or dataset
viii. Write your programme to your discretion

Find attached an example of codes used to process the satellite images. This process can be used to analyze more than one terabytes of data if the data has the same format.

//header files

```
#include "stdio.h"
#include "stdlib.h"
#include "TAxis.h"
#include "TH2D.h"
#include <TProfile.h>
#include "TF2.h"
#include <TF1.h>
#include "TMath.h"
#include "TCanvas.h"
#include "TStyle.h"
#include "TRandom3.h"
#include <TSystem.h>
#include "TVirtualFitter.h"
#include "TList.h"
#include <iostream>
#include "TArrayD.h"
#include <TApplication.h>
#include <TFitResult.h>
#include <fstream>
#include <TImage.h>
#include <TLegend.h>
#include <cmath>
#include <fstream>
#include <string>
#include <sstream>
#include <opencv2/core/core.hpp>
#include <opencv2/highgui/highgui.hpp>
#include <opencv2/imgproc/imgproc.hpp>
#include <algorithm>
#include <vector>
#include <stdio.h>
#include <stdlib.h>
#include <map>
#include <TApplication.h>
```

```cpp
#include <TRoot.h>
#include <TFile.h>
#include <TNtuple.h>
#include <TASImage.h>
#include <TImage.h>
#include <TTreeReader.h>
#include <TLegendEntry.h>
#include <TCutG.h>
#include <experimental/string_view>

using namespace std;
using namespace cv;
```

//Library linker
```cpp
R__ADD_LIBRARY_PATH (/usr/local/include/opencv2);
R__LOAD_LIBRARY(libopencv_highgui.2.4.13.dylib);
R__LOAD_LIBRARY(libopencv_core.2.4.13.dylib);
```

// Link to save results in ërootí file
```cpp
  ofstream myfile;
```

// Globals
```cpp
TObject *MyObj = 0;
double xfitRange1;
double xfitRange2;
double yfitRange1;
double yfitRange2;
double scale;   // height of peak
double width;   // width of peak
double bg;   // backgd
Double_t xpeak;
Double_t ypeak;
```

//Prototypes

```
TH2I* Remove_hot_pixels(TH2I* h);
void drawHist(TH2D* hist, TCanvas &c);
TF2* getFitFunction(string funcName);
Double_t get2DIntegral(TF2* fn, Double_t limit, Double_t stepSize);
Double_t gausFn(Double_t *x, Double_t *par) ;



int main(int argc, char **argv) {

  if(argc==1){
printf("Usage  : %s  <inputfile 1>   <rootfile> \n",argv[0]);
printf("Option : nothing yet \n");
exit(0);
  }
```

// **Extract input file name 1 from the argument list**

```
    char input_file_1[100];   // input file
    strcpy(input_file_1,argv[1]);
```

// **Extract ROOT output file name from the argument list**

```
    char root_file[64];   // root file
    strcpy(root_file,argv[2]);

    cout << "\n"<<input_file_1<<" and "<<root_file<<" \n\n";

    gROOT->Reset();
    TApplication*TheApp=new    TApplication("DAQanalisys",&argc,argv);
    TCanvas *MyCanvas = new TCanvas("canv", "Stuff",1200,800);
    MyCanvas->Divide(2,2);
```

// **Make a soft link in the project directory to the data called DATA**

```
    TString PathToData = "DATA/";
    TString input_1 = PathToData+input_file_1;
    std::cout << "Reading file  " << input_1 << endl;
```

```
char *y = new char[input_1.Length() + 1];
std::strcpy(y, input_1);

cv::Mat image_in;
    cv::Mat image_out;
image_in     =     cv::imread(y,     CV_LOAD_IMAGE_GRAYSCALE     |
CV_LOAD_IMAGE_ANYDEPTH);  // Read the file
delete[] y;
if(! image_in.data ) for invalid input
{
    cout <<  "Could not open or find the image" << std::endl ;
    return -1;
}


cout<<"Filling histo"<<endl;
Double_t * argb = new Double_t [image_in.rows*image_in.cols];
TH2I     *h2     =     new     TH2I("h2","h2",image_in.cols,0,image_in.cols-
1,image_in.rows,0,image_in.rows-1);
TH2I     *h2_1     =     new     TH2I("h2_1","h2_1",image_in.cols,0,image_in.cols-
1,image_in.rows,0,image_in.rows-1);
ushort pixel ;
for(int i = 0; i < image_in.cols; i++)
  {
  for(int j = 0; j < image_in.rows; j++)
   {
    pixel = (ushort)image_in.at<ushort>(j, i);
    argb[i*image_in.rows+j] = pixel;
    h2->Fill(i+1,image_in.rows-j,pixel);
            std::cout << "("<< i<<", "<<j<<") "<<"Pixels = " << pixel << endl;
   }
  }
UInt_t yPixels = image_in.rows;
UInt_t xPixels = image_in.cols;
```

```
std::cout << "xPixels = " << xPixels << " :  yPixels = " << yPixels << endl;


gStyle->SetCanvasPreferGL(kTRUE);
cout<<"Now plotting"<<endl;
MyCanvas->cd(1);
h2->SetStats(kFALSE);
h2->Draw("lego2");
h2_1 = Remove_hot_pixels(h2);
MyCanvas->cd(2);
h2_1->SetStats(kFALSE);
h2_1->Draw("lego2");


TH1D *h1;
h1 = h2_1->ProjectionX();
MyCanvas->cd(3);
h1->Draw("HIST");
MyCanvas->Update();
gPad->Modified();          // make sure
gPad->Update();             // hist is drawn
gSystem->ProcessEvents();


// Put errors on the Histogram to force the fit to respect the peak
Double_t h_Max = h2_1->GetBinContent(h2_1->GetMaximumBin());
cout <<  " Max value  " << h_Max << endl;
for (int row=0; row<xPixels; ++row) {
  for (int col=0; col<yPixels; ++col) {
    Double_t h_bin = h2_1->GetBinContent(h2_1->GetBin(row+1,col+1));
    if( h_bin < 0.1 * h_Max){
     h2_1->SetBinError(row+1,col+1,100);

     }
    else{
     h2_1->SetBinError(row+1,col+1,sqrt(h_bin));

     }
```

```
    }
  }


// Set the fitting parameters start value and fit range
xpeak = h1->GetXaxis()->GetBinCenter(h1->GetMaximumBin()); // x pos of
peak
h1 = h2_1->ProjectionY();
ypeak = h1->GetXaxis()->GetBinCenter(h1->GetMaximumBin()); // y pos of
peak
cout << "Peak position start values  (" << xpeak << " ,  " << ypeak <<")" <<
endl;
scale = 20000;   // height of peak
width = 20;   // width of peak
bg = 2000;   // background

Int_t plot_range = 100;
          Int_t a;
TH2I *h2_2 = new
  TH2I("h2_2","h2_2",2*plot_range+1,(Int_t)xpeak-
plot_range,(Int_t)xpeak+plot_range+1,
              2*plot_range+1,(Int_t)ypeak-
plot_range,(Int_t)ypeak+plot_range+1);
    for ( int i = 0; i < 2*plot_range+1; i++) {
    for ( int j = 0; j < 2*plot_range+1; j++) {
      a    =    h2_1->GetBinContent(i+(Int_t)xpeak-plot_range,j+(Int_t)ypeak-
plot_range);
      h2_2->Fill(i+(Int_t)xpeak-plot_range,j+(Int_t)ypeak-plot_range,a);

    }
  }
xfitRange1 = xpeak-100;
xfitRange2 = xpeak+100;
```

```
            yfitRange1 = ypeak-100;
            yfitRange2 = ypeak+100;
```

// Now draw Histo over smaller range to prep for fit display

```
            TCutG *cutg = new TCutG("cutg",5);
            cutg->SetVarX("x");
            cutg->SetVarY("y");
            double xplotRange1 = xpeak-50;
            double xplotRange2 = xpeak+50;
            double yplotRange1 = ypeak-50;
            double yplotRange2 = ypeak+50;
            cutg->SetPoint(0,xplotRange1,yplotRange1);
            cutg->SetPoint(1,xplotRange1,yplotRange2);
            cutg->SetPoint(2,xplotRange2,yplotRange2);
            cutg->SetPoint(3,xplotRange2,yplotRange1);
            cutg->SetPoint(4,xplotRange1,yplotRange1);
            MyCanvas->cd(4);
            h2_2->Draw("surf2");
            MyCanvas->Update();
            gPad->Modified();        // make sure
            gPad->Update();          // hist is drawn
            gSystem->ProcessEvents();
```

//Set the fit function to be any of the above

```
        cout<<"getting fit function"<<endl;
        TF2* fitFunc = getFitFunction("gaus2D");
        fitFunc->SetParameters(xpeak,ypeak, scale, width,bg);
        fitFunc->SetNpx(40);
        fitFunc->SetNpy(40);

        TFitResultPtr r = h2_2->Fit(fitFunc,"SN");


        // The output .root file
```

```
    TString filename_ROOT = "root_file";
    TFile *f = new TFile(filename_ROOT,"RECREATE");

    cout << endl<<"Results of fit"<<endl;
    cout << "Peak pos :  (" << r->Parameter(0) << " ,  " << r->Parameter(1) <<")" <<
endl;
    cout << "Scale    :  (" << r->Parameter(2) << endl;
    cout << "Width    :  (" << r->Parameter(3) << endl;

    fitFunc->SetParameters(r->Parameter(0),r->Parameter(1),r->Parameter(2),r-
>Parameter(3));

    fitFunc->Draw("surf1 same bb [cutg]");
    MyCanvas->Update();
    gPad->Modified();                    // make sure
    gPad->Update();            // hist is drawn
    gSystem->ProcessEvents();
    float calibration = 1; //pixels/mm
    float units = 1.;
    cout<<"FWHM="<<2.35*r>Parameter(3)*units/calibration<<endl;
char output_file[100];
    sprintf(output_file, "%s.txt",root_file);
    myfile.open (output_file, ios::app);

    f->Write();
    std::cout << "\n Done  ...  Now waiting...";
    std::cout << "\n ==> Double  click  mouse  button  in  graphics  window  to  end
program.\n\n\n";
    int n=1;
    while (n>0) {
      MyObj = gPad->WaitPrimitive();
            if (!MyObj) break;
            printf("Loop      i=%d,      found      objIsA=%s,      name=%s\n",n,MyObj-
>ClassName(),MyObj->GetName());
```

```
        }

    myfile.close();
    delete h2;
    delete h2_1;
    delete f;


cv::waitKey(0);                          // Wait for a keystroke in the window


}


// definition of the prototypes
TH2I* Remove_hot_pixels(TH2I* h1) {
  Int_t ncellsx = h1->GetNbinsX();
  Int_t ncellsy = h1->GetNbinsY();
  TH2I * h2 = (TH2I*)h1->Clone("h2");
  Int_t pixel;
  Int_t pixel_av;
  Int_t pixel_1;
  Int_t pixel_2;
  Int_t pixel_3;
  Int_t pixel_4;
  Int_t n_bad_pixels = 0;
  for (int row=5; row<ncellsx-5; ++row) {
    for (int col=5; col<ncellsy-5; ++col) {
      pixel = h1->GetBinContent(h1->GetBin(row+1,col+1));
      pixel_1 = h1->GetBinContent(h1->GetBin(row+1,col));
      pixel_2 = h1->GetBinContent(h1->GetBin(row+1,col+2));
      pixel_3 = h1->GetBinContent(h1->GetBin(row,col+1));
      pixel_4 = h1->GetBinContent(h1->GetBin(row+2,col+1));
      pixel_av = (pixel_1 + pixel_2 + pixel_3 + pixel_4)/4;
      if( pixel >   pixel_av +  5.0*sqrt(pixel_av) +  10  ||   pixel <   pixel_av -
5.0*sqrt(pixel_av) - 10 ){
        n_bad_pixels++;
```

```
      //cout << "Found hot pixel at :  " << row << " , " << col << endl;
      h2->SetBinContent(h1->GetBin(row+1,col+1), pixel_av);

    }
   }
  }
  cout << "Number of bad pixels = " << (float)n_bad_pixels/( (float)ncellsx *
(float)ncellsy ) << endl;
  return h2;
}


void drawHist(TH2D* hist, TCanvas &c) {
  hist->GetXaxis()->SetTitle("X [cm]");
  hist->GetYaxis()->SetTitle("Y [cm]");
  hist->GetZaxis()->SetTitle("Grey Scale");

  hist->GetXaxis()->SetTitleOffset(1.5);
  hist->GetYaxis()->SetTitleOffset(1.5);
  hist->GetZaxis()->SetTitleOffset(1.5);

  string name = hist->GetName();
  name += "c";
  c.cd();

  hist->Draw("lego2");
  c.Update();

}


Double_t gausFn(Double_t *x, Double_t *par) {
  Double_t xv = x[0] - par[0];
  Double_t yv = x[1] - par[1];
  Double_t r = TMath::Sqrt(xv * xv + yv * yv);
  return par[2] * TMath::Gaus(r, 0, par[3])+par[4];
}
```

```
TF2* getFitFunction(string funcName) {
    TF2* fitFunc = nullptr;

    if (funcName == "gaus2D") {
        fitFunc = new TF2("gaus2D", gausFn, xfitRange1, xfitRange2, yfitRange1,
yfitRange2, 5);
        fitFunc->SetParName(0, "x pos");
        fitFunc->SetParName(1, "y pos");
        fitFunc->SetParName(2, "scale");
        fitFunc->SetParName(3, "Sigma");
        fitFunc->SetParName(4, "backgd");
    }

    return fitFunc;
}



Double_t get2DIntegral(TF2* fn, Double_t limit, Double_t stepSize) {
    double integral = 0;
    for (double x = -limit; x < limit; x += stepSize) {
        for (double y = -limit; y < limit; y += stepSize) {
            if (x * x + y * y > limit) continue;
            integral += fn->Eval(x, y);
        }
    }
    return integral * (stepSize * stepSize);
}
```

The beauty of writing a compiled program is that results or outputs can be stored in predefined folders. The output files or images can be saved with their original file names or titles. Hence, it is easier to work with volume of data that is above one terabytes. Depending on the body of the program, it is interesting to note that the execution time is impressive. This means that a very fast computer may not be used to work on 'big data' as earlier proposed. The definition and scope of 'big data' in environmental science will be discussed in chapter five.

In writing macro, the library may be linked to the main program statically. Much header file may not be defined as in a program. Global parameters are defined within the 'void' or 'main' prototype. Find attach an example of a macro written to analyze large volume of dataset.

```
//Header file
#include "opencv2/highgui/highgui.hpp"
#include "opencv2/imgproc/imgproc.hpp"
#include <iostream>
#include <stdio.h>
#include <fstream>
using namespace std;
using namespace cv;

//Globals
Mat src, dst, des;

//prototype
char window_name[20]="Pixel Value Demo";

int main( int argc, char** argv )
{


    /// Load image from folder

    vector<String> filenames;

    string folder = "/Users/emetere/Desktop/processed/*.tif";
    cv::glob(folder, filenames);
```

```
for(size_t i = 0; i < filenames.size(); ++i)
{
    cv::Mat3b const src = imread(filenames[i], cv::IMREAD_COLOR);

    if( !src.data )
    { return -1; }
```

/// **Separate the image in 3 places (B, G and R)**
```
vector<Mat> bgr_planes;
split( src, bgr_planes );
```

/// **Establish the number of bins**
```
int histSize = 256;
```

```
/// Set the ranges ( for B,G,R) )
float range[] = { 0, 256 } ;
const float* histRange = { range };
```

```
bool uniform = true; bool accumulate = false;
```

```
Mat b_hist, g_hist, r_hist;
```

/// **Compute the histograms:**
```
    calcHist( &bgr_planes[0], 1, 0, Mat(), b_hist, 1, &histSize, &histRange, uniform,
accumulate );
    calcHist( &bgr_planes[1], 1, 0, Mat(), g_hist, 1, &histSize, &histRange, uniform,
accumulate );
    calcHist( &bgr_planes[2], 1, 0, Mat(), r_hist, 1, &histSize, &histRange, uniform,
accumulate );
```

// **Draw the histograms for B, G and R**
```
int hist_w = 512; int hist_h = 400;
int bin_w = cvRound( (double) hist_w/histSize );
```

```
Mat histImage( hist_h, hist_w, CV_8UC3, Scalar( 255,255,255) );
```

/// Normalize the result to [0, histImage.rows]

```
normalize(b_hist, b_hist, 0, histImage.rows, NORM_MINMAX, -1, Mat() );
normalize(g_hist, g_hist, 0, histImage.rows, NORM_MINMAX, -1, Mat() );
normalize(r_hist, r_hist, 0, histImage.rows, NORM_MINMAX, -1, Mat() );
```

/// Draw for each channel

```
for( int i = 1; i < histSize; i++ )
{
    line( histImage, Point( bin_w*(i-1), hist_h - cvRound(b_hist.at<float>(i-1)) ) ,
        Point( bin_w*(i), hist_h - cvRound(b_hist.at<float>(i)) ),
        Scalar( 255, 0, 0), 2, 8, 0 );
    line( histImage, Point( bin_w*(i-1), hist_h - cvRound(g_hist.at<float>(i-1)) ) ,
        Point( bin_w*(i), hist_h - cvRound(g_hist.at<float>(i)) ),
        Scalar( 0, 255, 0), 2, 8, 0 );
    line( histImage, Point( bin_w*(i-1), hist_h - cvRound(r_hist.at<float>(i-1)) ) ,
        Point( bin_w*(i), hist_h - cvRound(r_hist.at<float>(i)) ),
        Scalar( 0, 0, 255), 2, 8, 0  );
}
```

/// Display

```
namedWindow("calcHist Demo", CV_WINDOW_AUTOSIZE );
imshow("calcHist Demo", histImage );
histImage.convertTo(dst, CV_8UC3);
std::ostringstream name;
name  <<  "/Users/emetere/Desktop/intensity/jpg/Phi-1to4speed0.01_2017-10-
21-201727-0" << i << ".jpg";
cv::imwrite(name.str(),dst);
```

///Saving into text

```
    int j, k;
    std::ofstream pfout("/Users/emetere/Desktop/intensity/output_test.txt");
    Vec3b pix=src.at<Vec3b>(k,j);
```

```
        int Red=src.at<cv::Vec3b>(k,j)[2];
        int Green= src.at<cv::Vec3b>(k,j)[1];
        int Blue = src.at<cv::Vec3b>(k,j)[0];
        int y= src.at<cv::Vec3b>(k,j)[3];
        int x = src.at<cv::Vec3b>(k,j)[4];
        int Totalintensity = 0;
        for (int u=0; u < src.rows; ++u){
        for (int v=0; v < src.cols; ++v){
           Totalintensity += (int)src.at<uchar>(u, v);
             }
           }

        // Find avg lum of frame
        float avgLum = 0;
        avgLum = Totalintensity/(src.rows * src.cols);

        cout<<"  x= "<<x<<"  y= "<<y<<"  Red= "<<Red<<"  Green=  "<<Green<<"
Blue= "<<Blue<<"  intensity of images = "<<avgLum<<" \t\n";
        pfout<<"  x= "<<x<<"  y= "<<y<<"  Red= "<<Red<<"  Green=  "<<Green<<"
Blue= "<<Blue<<"  intensity of images = "<<avgLum<<" \t\n";

    ///geting axis
    Vec3f intensity = src.at<Vec3f>(y, x);
    float blue = intensity.val[0];
    float green = intensity.val[1];
    float red = intensity.val[2];
    pfout<<"  Red= "<<red<<"  Green=  "<<green<<"  Blue= "<<blue<<" \t\n";
    cout<<"  Red= "<<red<<"  Green=  "<<green<<"  Blue= "<<blue<<" \t\n";

    ///plot function
    Mat plot_result;

}
waitKey(0);
    return 0;
}
```

4.4   Planning a Computational and Mathematical Model Design Insights into 'Big Data' Analysis

Planning a computational or Mathematical Model large volumes of dataset of images require the following skill:

i.   an in-depth understanding of a code to extract required information from the image;
ii. quantifying the information to numbers;
iii. understand the theoretical background of quantified information;
iv. be able to manipulate the information to suite your research objectives.

These skills are essential to dig-out unimaginable information from an image and apply it to a very large dataset (i.e. big data) of images of same kind. The main objective of this section is to show how to generate a computational model from a large dataset of images.

In this illustration, 1568 images of a dataset were analyzed. The matrix of each images were categorized into column (y-axis) pixels and row (x-axis) pixels. Based on the listed information, it is possible to expand or manipulate numerical information extracted from the dataset. In Fig. 4.56a, b, the mean-intensity of the images was extracted. The polynomial of the trend is shown at the top-left corner of the figure. The values of the polynomial trend may be used to develop a mathematical model. However, the focus is to show how computational model can be developed from large dataset. Hence, the steps that shall be considered in the computational model are:

i.   give a 2D-plot of the images in sequential order. In this book, the mean-pixel of each images were considered;
ii. choose the significant spots of interest. This process specifies on unique information that may be indicated as new or existing concept;
iii. do the 3D-plot of each spot;
iv. do a Gaussian fit for each spot to know if they all have same set of principle;
v.   consider more parameters with emphasis on the spot analysis;
vi. draw out your conclusion on the spot analysis.

It is observed from Fig. 4.56a, b that significant perturbation of the linearity of the graph was at 550, 750, 825, 910, 1020, 1100 and 1125. This means that the information on these spots may refer to the already known principle or may refer to a new event. The easiest way of knowing if information exists or is a new concept is your familiarity to the literature of the subject-matter. The spot analysis was done using a 3D image as in Fig. 4.57. To understand the significant changes that has occurred in each spot, a spot will be taken from the linear or uniform trend of the graph. Hence, Fig. 4.57a is the normal spot and Fig. 4.57b-h refers to the highlighted spots listed above (i.e. 550, 750, 825, 910, 1020, 1100 and 1125).

As mentioned above, the relevance of the spot analysis is to critically examine if a new result has been found or an existing result has been validated. For example, the Gwydion software was used to analyze the satellite map as shown below.

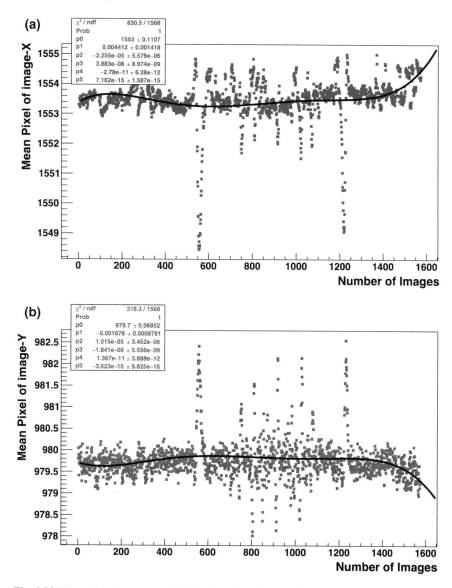

Fig. 4.56 Mean pixel of images. **a** Pixel in the column-imageY, **b** pixel in the row-imageX

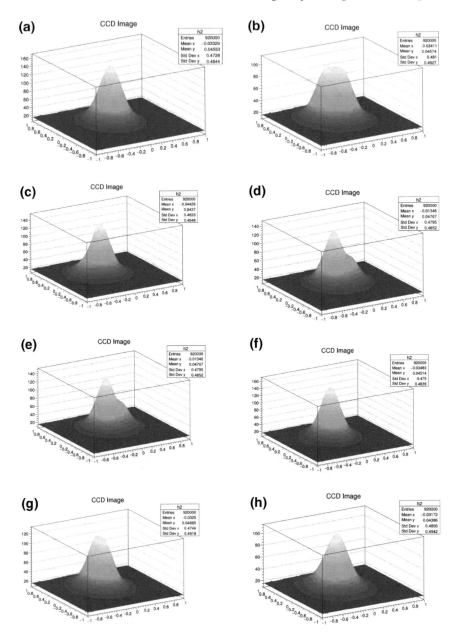

Fig. 4.57 The 3D-plot for each spot. **a** Normal spot, **b** 550, **c** 750, **d** 825, **e** 910, **f** 1020, **g** 1100, **h** 1125

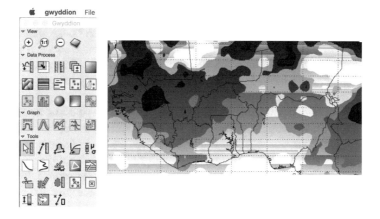

The above image can be used to design specific objects in 2D as shown below. It is interesting that the 2D image can be further processed as a whole (as shown below) or individually processed as shown in Fig. 4.57. Each of the points can be further separated either by snapping areas of interest the use of crop tools. It was observed that the analysis of each spot is not the same when further processed as shown in Fig. 4.57. This process is tedious, however, through the use of openCV, it is possible to analyze large dataset without much stress.

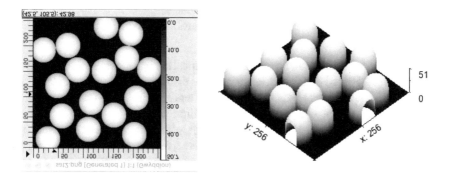

Taking a keen look at the 3D-plot (i.e. comparing Fig. 4.57b–h with Fig. 4.57a), it was observed that at the 555th image, the peak is flat and broadened (Fig. 4.57b). Further interpretation depends on the readers' familiarity with literature. There is an elongation on the left-hand side of the peak in the 750th image (Fig. 4.57c). In practical terms, there must be a significant transition from Fig. 4.57b, c. The quantification of the transition can be done by considering the x-projection and y-projections of both graphs. Recall the mathematical significance of the x- and y-projections has been discussed in the early part of this chapter. Figure 4.57d, e show the event in the spot-analysis of the 825th and 910th images. Both images have a deformed and elongated right-hand side of the peak. This certainly means that the same set of event occurred on each spot.

Figure 4.57f show a significant deformation at the base of the peak. This means that there is a significant event between the 910th and 1020th image. Aside the use of the x- and y-projections, a code that calculates the difference between the pixels of each image can be written as shown below in Sect. 4.4.1. Figure 4.57g show a flattened peak i.e. similar to Fig. 4.57b. Hence, the 555th may have same event as the 1100th image.

Figure 4.57h has a significant broadening at the base of the peak and a very narrow peak at the top. This event is almost similar to Fig. 4.57f. In general, the spot analysis reveal that events can alternatively occur within the confine of same experiment. Hence the next step was to apply the Gaussian fit as shown in Fig. 4.58a–h. It is observed that Gaussian fit for Fig. 4.58g was not appropriate. Hence, the need to further statistically investigate the whole images by looking at the kurtosis (Fig. 4.59a, b), deviation error (Fig. 4.60a, b) and skew (Fig. 4.61a, b).

The surface appearance of Fig. 4.59a, b may look somewhat similar but a keen or closer observation shows that there are significant differences between both images. First, the polynomial factors in both images differ. Second, there are additional spots over the main curve in Fig. 4.59b. So what does this signify? The interpretation of the Fig. 4.59 should be related to the statistical significance of kurtosis. Kutosis is defined as a measure of the "tailedness" of the probability distribution of a real-valued random variable. It is a descriptor of the shape of a probability distribution. Hence, kurtosis can be interpreted based on the following:

i.   scaled version of the fourth moment of the dataset;
ii. the type of kurtosis distribution i.e. mesokurtic, leptokurtic and leptokurtic;
iii. scaled version of the fourth L-moment.

The details of kurtosis can be obtained from most statistical textbooks. The distribution of Fig. 4.59a, b is referred to as platykurtic distribution because its distribution is less than 3. These types of distributions have slender tails, broad-looking peak and a peak that is smaller than a mesokurtic distribution.

Figure 4.60a, b describes the deviation error within the images. The trends of Fig. 4.60 are more similar than Fig. 4.59. However, there are differences looking at the values of the polynomial factors that is located at the top-left corner of the graph. Secondly, the mean pixel along the y-axis of the images (Fig. 4.60a) is lower than the mean pixel along the x-axis (Fig. 4.60b). This shows that there are major changes along the column or latitude of the images. Like the kurtosis, the interpretation of Fig. 4.60 is based on the statistical definition of deviation error. Deviation error or standard error of the dataset is defined as the standard deviation of its sampling distribution. Reader should note that there is a difference between standard deviation and deviation/standard error. Standard deviation is a measure of dispersion of the data from the mean while deviation error is a measure of how precise is the estimate of the mean. In general, the deviation in Fig. 4.60 shows how similar the images are.

Figure 4.61a, b differ from Figs. 4.59 and 4.60 because the skewness of image-X is significantly different from the skew of image-Y. This shows that the skewness of the image pixels on the column and rows gives different information. The question is what is skewness? Skewness is defined as the measure of the asymmetry of the probability

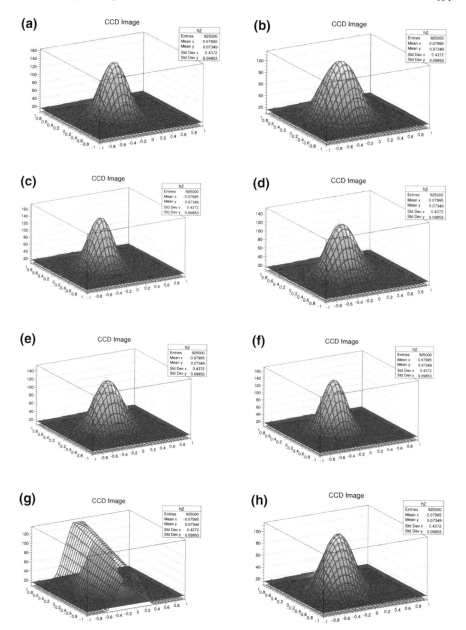

Fig. 4.58   The Gaussian fit for each spot. **a** Normal spot, **b** 550, **c** 750, **d** 825, **e** 910, **f** 1020, **g** 1100, **h** 1125

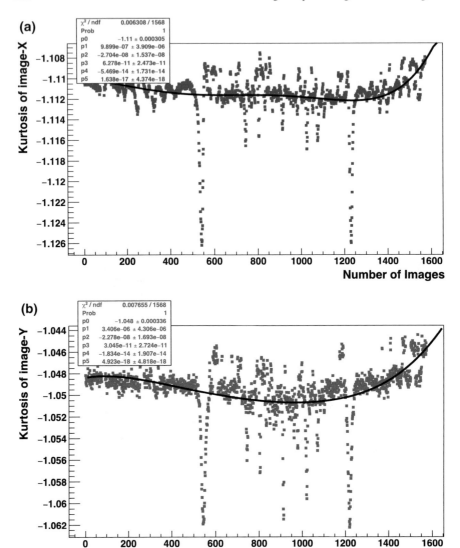

Fig. 4.59 Kurtosis of images. **a** Pixel in the column—imageY. **b** Pixel in the row—imageX

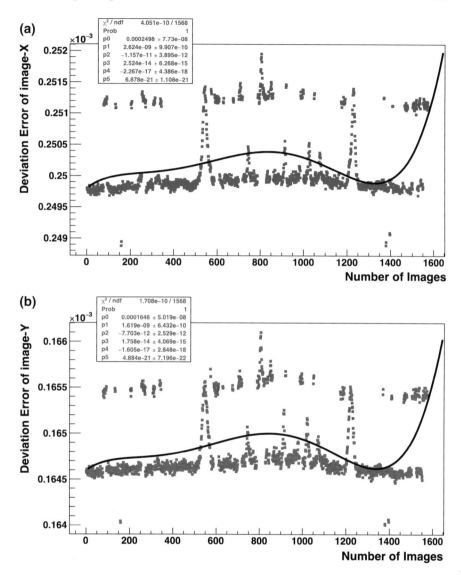

Fig. 4.60 Deviation error of images. **a** Pixel in the column—imageY. **b** Pixel in the row—imageX

distribution of a real-valued random variable about its mean. The Pearson's first coefficient is the skewness of the dataset. It is measured by subtracting the mean from the mode, and then divide the difference by the standard deviation of the data. Unlike kurtosis, the qualitative interpretation of the skew is complicated and unintuitive. With respect to Fig. 4.61, the positive skew (right-skew) indicates that the tail on the right side is longer or fatter than the left side while the negative skew (left-skew) indicates a long tail in the negative direction on the number line.

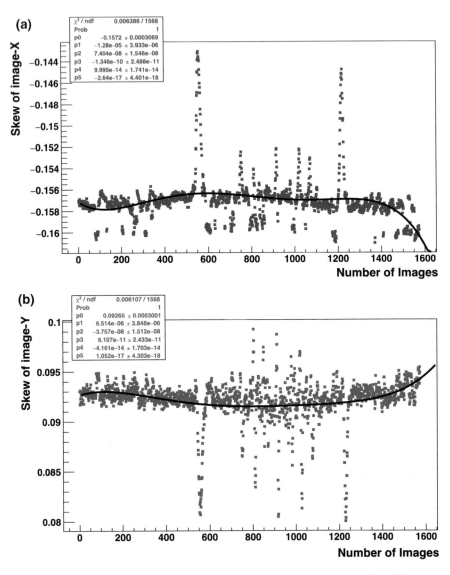

Fig. 4.61 Skew of images. **a** Pixel in the column—imageY. **b** Pixel in the row—imageX

The second part of this section is the development of a mathematical model. Let us assume a general polynomial for the trend of the images as:

$$z_x = p_5x^5 + p_4x^4 + p_3x^3 + p_2x^2 + p_1x + p_0 \qquad (4.5)$$

$$z_y = p_5y^5 + p_4y^4 + p_3y^3 + p_2y^2 + p_1y + p_0 \qquad (4.6)$$

Here z_x the pixel of image-X, z_y is the pixel of image-Y, x and y represents the individual components of the pixel in x- and y-axis. The third dimension is the introduction of image resolution (z_z).

$$z_z = \frac{\text{Number of pixel } (z)}{\text{Field of view (fov)}} \tag{4.7}$$

if the z is written as:

$$z = \sum x \times \sum y = \sum yx \tag{4.8}$$

$$z_z = \frac{1}{\text{fov}} \sum yx \approx \frac{1}{\text{fov}} xy \tag{4.9}$$

$$\text{fov} = \alpha \frac{D}{d} \tag{4.10}$$

where D is dimension of full image (xy), d is the dimension of target (≈ 1), α is the targets angular extent (≈ 1). The target parameters were set to unity since no target was considered.

If Eqs. 4.5, 4.6 and 4.9 are differentiated with respect to time (t).

$$\frac{dz_x}{dt} = vp_5 x^4 + vp_4 x^3 + vp_3 x^2 + vp_2 x + vp_1 \tag{4.11}$$

$$\frac{dz_y}{dt} = \vartheta p_5 y^4 + \vartheta p_4 y^3 + \vartheta p_3 y^2 + \vartheta p_2 y + \vartheta p_1 \tag{4.12}$$

$$\frac{dz_z}{dt} = \vartheta xz \tag{4.13}$$

here $v = \frac{dx}{dt}$ and $\vartheta = \frac{dy}{dt}$.

Then z must be introduced to Eqs. 4.11 and 4.12. Thus multiply Eq. 4.11 by y and Eq. 4.12 by x.

$$y \frac{dz_x}{dt} = yvp_5 x^4 + yvp_4 x^3 + yvp_3 x^2 + yvp_2 x + yvp_1 \tag{4.14}$$

$$x \frac{dz_y}{dt} = x\vartheta p_5 y^4 + x\vartheta p_4 y^3 + x\vartheta p_3 y^2 + x\vartheta p_2 y + x\vartheta p_1 \tag{4.15}$$

$$\frac{dz_z}{dt} = \vartheta xz \tag{4.16}$$

Assume that $x^n x^m = 0$

$$y \frac{dz_x}{dt} = vp_2 z + vp_1 y \tag{4.17}$$

Table 4.2 Numerical
solutions of governing
equations

t	z_x	z_y	z_z
0	10	1	1
0.1	10	0.999997	1.0202
0.55	10	0.999981	1.11628
1.9242	10.0002	0.999935	1.46938
3.29839	10.0004	0.999889	1.93419
4.67259	10.0006	0.999843	2.54605
6.04679	10.0009	0.999797	3.35148
7.42098	10.0013	0.999751	4.41175
8.79518	10.0019	0.999706	5.80752
10.1694	10.0026	0.999661	7.64502
11.5436	10.0035	0.999617	10.0641
12.9178	10.0048	0.999573	13.2491
14.292	10.0065	0.999531	17.4428
15.6662	10.0086	0.999489	22.965
17.0404	10.0115	0.99945	30.2376
18.4146	10.0153	0.999413	39.8168
19.7888	10.0203	0.999379	52.4371
21.163	10.0268	0.999349	69.0684
22.5371	10.0355	0.999324	90.9935
23.9113	10.0469	0.999306	119.911
25.2855	10.0619	0.999297	158.076
26.6597	10.0817	0.999301	208.487
28.0339	10.1079	0.99932	275.148
29.4081	10.1424	0.99936	363.422
30.7823	10.188	0.999429	480.542
32.0039	10.2408	0.999521	616.737
33.2255	10.3086	0.999651	792.692
34.4471	10.3959	0.999829	1020.77
35.6686	10.5083	1.00007	1317.66
36.8902	10.6538	1.0004	1706.24
38.1118	10.8425	1.00083	2218.4
39.3334	11.0884	1.00141	2899.54
40.382	11.3598	1.00206	3668.74
41.4307	11.7043	1.0029	4671.92
42.4794	12.1448	1.00397	5998.29
43.408	12.6404	1.00519	7549.29
44.3366	13.2674	1.00674	9600.36
45	13.8196	1.00812	11489.1

$$x\frac{dz_y}{dt} = \vartheta\, p_2 z + \vartheta\, p_1 x \tag{4.18}$$

$$\frac{dz_z}{dt} = \vartheta\, xz \tag{4.19}$$

Hence, the three governing equations are 4.17, 4.18 and 4.19. The governing equation was solved using the numerical analysis of C++ Odeint library that can be obtained at Odient (2018). Odeint is a modern C++ library for numerically solving Ordinary Differential Equations. The solution of Eqs. 4.17–4.19 when $\vartheta = 0.02$, v $= 0.018$, t $= 45$ s is given in Table 4.2. While the resolution (z_z) of the images is expected to increase significantly in 45 s, The values of the mean pixel increases at the x-axis at higher ratio than the mean pixel of the image at the y-axis. What is the significance of this result? The selected spots (i.e. 550, 750, 825, 910, 1020, 1100 and 1125) are spot of highest resolution. Within the high resolution (i.e. spots) the value for image-X is expected to be higher than image-Y. These results are validated by Figs. 4.56, 4.59, 4.60 and 4.61 were the values of image-X is always higher than image-Y.

Therefore, the mathematical model has exhibited high sensitivity to be used to test other results.

4.5 Macro to Determine the Difference of Images by Pixel

The code to determine the difference between two images—based on its pixel content can be itemized as:

i. write the header files
ii. state the main program header
iii. load the images
iv. resize both images to have same dimensions
v. define the height and width
vi. find the pixels along the height and width
vii. calculate the percentage change between the images

The code is written in C++, however, the reader can use any familiar computer language to write the same code.

```cpp
#include "opencv2/highgui/highgui.hpp"
#include "opencv2/imgproc/imgproc.hpp"
#include "opencv2/core/core.hpp"
#include <iostream>
#include <algorithm>
#include <vector>
#include <stdio.h>
#include <stdlib.h>
#include <fstream>

using namespace cv;
using namespace std;


int main(){

    auto firstImage =
imread("/Users/emetere/Desktop/satellite/sat001.png",CV_LOAD_IMAGE_ANYCOL
OR);
    auto secondImage = imread("/Users/emetere/Desktop/ satellite/sat002.png
",CV_LOAD_IMAGE_ANYCOLOR);
  cv::resize(secondImage, secondImage, firstImage.size());

  if (firstImage.size()==secondImage.size())
  {
     double totaldiff = 0.0 ; //holds the number of different pixels
     int h = firstImage.size().width;
     int w = firstImage.size().height;
     int hsecond = secondImage.size().height;
     int wsecond = secondImage.size().width;
     if ( w != wsecond || h != hsecond ) {
        std::cerr << "Error, pictures must have identical dimensions!\n" ;
        return 2 ;
     }

     for ( int y = 0 ; y < h ; y++ ) {
        uint *firstLine = ( uint* )firstImage.ptr<uchar>(y) ;
        uint *secondLine = ( uint* )secondImage.ptr<uchar>(y) ;
        for ( int x = 0 ; x < w ; x++ ) {
           uint pixelFirst = firstLine[ x ] ;
           double rFirst = ((double) cv::countNonZero(pixelFirst));
           double gFirst = ((double) cv::countNonZero(pixelFirst));
           double bFirst = ((double) cv::countNonZero(pixelFirst));
           //cout<<rFirst<<" "<<gFirst<<" "<<bFirst<<" "<<endl;
```

```
                    uint pixelSecond = secondLine[ x ] ;
                    double rSecond = ((double) cv::countNonZero(pixelSecond ));
                    double gSecond = ((double) cv::countNonZero(pixelSecond ));
                    double bSecond = ((double) cv::countNonZero(pixelSecond ));
                    totaldiff += std::abs( rFirst - rSecond ) / 255.0 ;
                    totaldiff += std::abs( gFirst - gSecond ) / 255.0 ;
                    totaldiff += std::abs( bFirst -bSecond ) / 255.0 ;


              }
           }
           std::cout << "The difference of the two pictures is " <<
           (totaldiff * 100)  / (w * h * 3)  << " % !\n" ;
        }
        else
        {
           cout<<"The dimension is wrong"<<endl;
        }

        waitKey(0);
        return 0;
   }
```

References

Acker, J. G., & Leptoukh, G. (2007). Online analysis enhances use of NASA earth science data. *EOS, 88,* 14–17.

Bond, T. C., & Bergstrom, R. W. (2006). Light absorption by carbonaceous particles: an investigative review. *Aerosol Science and Technology, 40,* 27–67.

Bond, T. C., Doherty, S. J., Fahey, D. W., Forster, P. M., Berntsen, T., DeAngelo, B. J., et al. (2013). Bounding the role of black carbon in the climate system: A scientific assessment. *Journal of Geophysical Research, 118*(11), 5380–5552.

Campbell, J. B. (2002). *Introduction to remote sensing.* New York London: The Guilford Press.

Emetere, M. E. (2016a). Statistical examination of the aerosols loading over Mubi-Nigeria: The satellite observation analysis. *Geographica Panonica, 20*(1), 42–50.

Emetere, M. E. (2016b). *Numerical modelling of West Africa regional scale aerosol dispersion.* Thesis submitted to Covenant University.

Emetere, M. E. (2017). Investigations on aerosols transport over micro- and macro-scale settings of West Africa. *Environ. Eng. Res., 22*(1), 75–86.

Emetere, M. E., & Akinyemi, M. L. (2017). Documentation of atmospheric constants over Niamey, Niger: a theoretical aid for measuring instruments. *Meteorological Applications, 24*(2), 260–267.

Emetere, M. E., Akinyemi, M. L., & Akinojo, O. (2015a). Parametric retrieval model for estimating aerosol size distribution via the AERONET, LAGOS station. *Environmental Pollution, 207*(C), 381–390.

Emetere, M. E., Akinyemi, M. L., & Akin-Ojo, O. (2015b). Aerosol optical depth pollution in selected areas trends over different regions of Nigeria: Thirteen years analysis. *Modern Applied Science., 9*(9), 267–279.

Emetere, M. E., Esisio, F., Oladapo, F. (2017b). Satellite observation analysis of aerosols loading effect over Monrovia-Liberia. *Journal of Physics: Conference Series, 852*(1), art. no. 012009. DOI: 10.1088/1742-6596/852/1/012009.

Emetere, M. E., Sanni, S. E., Emetere, J. M., & Uno, U. E. (2017a). Thermal Infrared remote sensing of hydrocarbon in Lagos-Southern Nigeria: Application of the thermographic model. *International Geomate Journal, 13*(39), 33–45.

Emetere, M. E., Sanni, S. E., & Tunji-Olayeni, P. (2017b). Atmospheric configurations of aerosols loading and retention over Bolgatanga-Ghana. *Journal of Physics: Conference Series, 852*(1), art. no. 012007. DOI: 10.1088/1742-6596/852/1/012007.

GeoEye. (2018). *GeoEye-1 satellite sensor (0.46 m).* https://www.satimagingcorp.com/satellite-sensors/geoeye-1/. Accessed January 12, 2018.

Koike, M., Moteki, N., Khatri, P., Takamura, T., Takegawa, N., Kondo, Y., et al. (2014). Case study of absorption aerosol optical depth closure of black carbon over the East China Sea. *Journal of Geophysical Research, 119*(1), 122–136.

Lacis, A. A., & Mishchenko, M. I. (1995). Climate forcing, climate sensitivity, and climate response: A radiative modeling perspective on atmospheric aerosols. In R. Charlson & J. Heintzenberg (Eds.), *Aerosol forcing of climate* (pp. 11–42). New York: John Wiley.

Landinfo. (2018). *Satellite imagery resolution comparison.* http://www.landinfo.com/GalSatResComp.htm. Accessed January 12, 2018.

Lindén, J., Thorsson, S., Boman, R., Holmer, B. (2012). *Urban climate and air pollution in Ouagadougou, Burkina Faso: An overview of results from five field studies* (pp. 1–88). University of Gothenburg. http://hdl.handle.net/2077/34289.

NASA. (2015). *NASA satellite camera provides "EPIC" view of earth.* https://www.nasa.gov/press-release/nasa-satellite-camera-provides-epic-view-of-earth. Accessed January 4, 2018.

Odient. (2018). Odeint solving ODE's in C++. http://headmyshoulder.github.io/odeint-v2/. Accessed February 17, 2018.

Omotosho, T. V., Emetere, M. E., & Arase, O. S. (2015). Mathematical projections of air pollutants effects over Niger Delta region using remotely sensed satellite data. *International Journal of Applied Environmental Sciences, 10*(2), 651–664.

Rafferty, J.P. (2010). *Storms, violent winds, and earth's atmosphere.* The Rosen Publishing Group (p. 95). SBN-13: 978-1615301140, ISBN-10: 1615301143.

Senghor, H., Machu, É., Hourdin, F., Gaye, A. T. (2017). Seasonal cycle of desert aerosols in western Africa: analysis of the coastal transition with passive and active sensors. *Atmospheric Chemistry and Physics, 17*, 8395–8410.

Tegen, I., Peter, H., Mian, C., Fung, I., Jacob, D, Penn, J. (1997). Contribution of different aerosol species to the global aerosol extinction optical thickness: Estimates from model results. *Journal of Geophysical Research, 102*(D20), 23895–23915.

Tirabassi, T., Moreira, D. M., Vilhena, M. T., da Costa, C. P. (2010). Comparison between Non-Gaussian puff model and a model based on a time-dependent solution of advection-diffusion equation. *Journal of Environmental Protection, 1*, 172–178.

Walcek, C.J. (2004). *A Gaussian dispersion/plume model explicitly accounting for wind shear.* https://ams.confex.com/ams/pdfpapers/79742.pdf. Accessed January 9, 2018.

Chapter 5
Big Data and Further Analysis

'Big Data' is a relative term used to describe a tremendously large data. The large data is inclusive of audio, video, unstructured text, social media information, and so much more. Its concept has gained wide publicity or attention in many disciplines. Interestingly, 'Big data' means different things to various disciplines. For example, in atmospheric study, 'big data' means volume of data as large as one terabytes and above. Meanwhile in particle physics, 'big data' is in petabytes and above. For communication outfit, 'big data' may mean zettabytes. Hence, there is the need for disciplinary and multi-disciplinary outfit or research institutes to embrace 'big data' technologies such as in-memory technologies, sensory (Internet of Things) equipment, Cloud Data Storage, magnetic storage, Big Data databases (e.g. MongoDB) etc.

The origin of 'big data' was traced to John Graunt (1663)—a statistical analyst who worked on a large volume of information on bubonic plague in Europe (Foote 2017). Herman Hollerith in 1881 worked with big data in U.S. Census Bureau to create the first big data analysis instrument—Hollerith Tabulating Machine. In other words, the concept of 'big data' has been used in form of machines, processes or applications in the past. In 2005, Roger Mougalas came-up with the word 'big data'. Since its identification and universal acceptability, universities, businesses and governments alike began to establish big data projects. Among application of 'big data', but not limited to the following are: in understanding and targeting consumers; self-optimization; improvement of healthcare; security and law enforcement improvement (Cleverism 2018); the music industry replaces intuition with Big Data studies (Dataversity 2018); being used by cybersecurity to stop cybercrime; to explore the universe using satellite exploration; establishing detailed research outcome such as association rule learning, classification tree analysis, genetic algorithms, machine learning, regression analysis, statistical analysis and social network analysis (Stephenson 2013).

The analytics techniques or methods for understanding big data are many according to specializations and disciplines. Some of analytics techniques that provide the most value for analyzing big data are visualization models, logistic regression, text analytics e.g. information extraction, audio analytics–speech analytics, video content

© Springer Nature Switzerland AG 2019
M. E. Emetere, *Environmental Modeling Using Satellite Imaging and Dataset Re-processing*, Studies in Big Data 54,
https://doi.org/10.1007/978-3-030-13405-1_5

analysis (VCA). Businesses use the visualization models such as Domo, Qlik, Tableau, Sisense, Reltio etc. However, there are some challenges associated with the big data concept. Borne (2014) identified the challenges of big data in an article titled "*top ten big data challenges—a serious look at 10 big data V's*" as: volume of data; variety of data; velocity of data within systems; veracity of data to ascertain its sufficiency to solve problems; validity of the source of data; value of data to specific disciplines; variability of data dynamism; venue showing the destination of data; vocabulary to classify data; and vagueness of data not qualified to be termed 'big data'. Qubole (2008) listed the challenges of big data (facing professionals in different fields of endeavor) as: scalability, lack of talent, actionable insights, data quality, security of data and cost management.

In the next section, the analytics of the large data obtained from MISR for one hundred and thirty locations over the region discussed in the previous chapter was discussed. The visualization model or Computer technique that was used was based on the CERN Root C++ open source.

5.1 Description of Data Source

The dataset was obtained from the Multi-angle Imaging SpectroRadiometer (MISR). As mentioned in chapter one, MISR operates at various directions, that is, nine different angles (70.5°, 60°, 45.6°, 26.1°, 0°, 26.1°, 45.6°, 60°, 20.5°) and gathers data in four different spectral bands (blue, green, red, and near-infrared) of the solar spectrum. The blue band is at wavelength 443 nm, the green band is at wavelength 555 nm, the red band wavelength 670 nm and the infrared band is at wavelength 865 nm. The blue band is used to analyse ice, snow, soil or water. The green band is to analyse Bathymetric mapping and estimating peak vegetation. The red band analyses the variable vegetation slopes and the infrared band analyses the biomass content and shorelines.

Thirteen years dataset was obtained for each of the one hundred and thirty locations across West Africa (see Fig. 5.1). The locations are Benin (Bohicon), Benin (Cotonou), Benin (Kandi), Benin (Natitingou), Benin (Parakou), Benin (Portnovo), Benin (Save), Burkina Faso (Banfora),Burkina Faso (Bobodiolasso), Burkina Faso (Kongoussi), Burkina Faso (Ouagadougou), Burkina Faso (Ouahigouya), Burkina Faso (Dori), Cameroun (Bamenda), Cameroun (Douala), Cameroun (Ebolowa), Cameroun (Garoua), Cameroun (Kousseri), Cameroun (Kumbo), Cameroun (Ngoundere), Cameroun (Younde), Capeverde (Assomada), Capeverde (Ponta), Capeverde (Praia), Chad (Abeche), Chad (Faya), Chad (Mao),Chad (Moundou), Chad (Ndjamena), Chad (Sahr), Cote d'Ivoire (Bondougou), Cote d'Ivoire (Sanpedro), Cote d'Ivoire (Daloa), Cote d'Ivoire (korhogo), Equitorial guinea (Bata), Equitorial guinea (Ebebiyin), Equitorial guinea (Malabo), Gambia (Basse), Gambia (Brikama), Gambia (Farafenni), Gambia (Serekunda), Guinea Bussau (Bafata), Guinea Bussau (Bussau), Guinea Bussau (Gabu), Ghana (Accra), Ghana (Bawku), Ghana (Bolgatanga), Ghana (Sunyani), Ghana (Takoradi), Ghana (Tamale), Guinea

Fig. 5.1 Study area: West Africa and its environ

(Conakry), Guinea (Koundara), Guinea (Macenta), Guinea (Nzerekore), Guinea (Siguiri), Guinea (kankan), Liberia (Buchanan), Liberia (Harper), Liberia (Monronvia), Liberia (Voinjama), Liberia (Yekepa), Mali (Mopti), Mali (Bamako), Mali (Gao), Mali (Kidal), Mali (Nioro), Mali (Segou), Mali (Sikasso), Mauritania (Aqjawajat), Mauritania (Kifah), Mauritania (Nawadibu), Mauritania (Nawaksut), Mauritania (Silibabi), Mauritania (Walatah), Mauritania (Zuwarat), Niger (Agadez), Niger (Arlit), Niger (Gaya), Niger (Magaria), Niger (Niamey), Niger (Tahoua), Nigeria (Abeokuta), Nigeria (Abuja), Nigeria (Calabar), Nigeria (Damaturu), Nigeria (Enugu), Nigeria (Gusau), Nigeria (Ibadan), Nigeria (Ikot Ekpene), Nigeria (Ilorin), Nigeria (Jos), Nigeria (Kaduna), Nigeria (Kano), Nigeria (Kastina), Nigeria (Lagos), Nigeria (Minna), Nigeria (Mubi), Nigeria (Ogbomoso), Nigeria (Ondo), Nigeria (Oshogbo), Nigeria (Sokoto), Nigeria (Onitsha), Nigeria (Owerri), Nigeria (Warri), Senegal (Dakar), Senegal (Louga), Senegal (Tambakounda), Senegal (Ziguinchor), Sierra Leone (Makeni), Sierra Leone (Binkolo), Sierra Leone (Kabala), Sierra Leone (Talama), Togo (Atakpame), Togo (Dapaong), Togo (Kara), Togo (Lome) and Togo (Sokode).

5.2 Data Analysis: Relevant Connection to Imagery

In this section, the discussion is based on specific location of interest as depicted in Chap. 4. The format of the plot is as follows: the first subsection tagged 'a' show the 3D image of AOD at 550 nm against the mean AOD (of the blue, green, red and infra-red band); the second subsection tagged 'b' show the 3D image of AOD 865 nm against Mean AOD; the third subsection tagged 'c' show the 2D image of AOD 440 nm against sum AOD (550, 670, 865 nm); the fourth subsection tagged 'd' show the 2D image of AOD 550 nm against sum AOD (440, 670, 865 nm); the fifth subsection tagged 'e' show the 2D image of AOD 670 nm against sum AOD (440, 550, 865 nm); the sixth subsection tagged 'f' show the 2D image of AOD 865 nm against sum AOD (440, 550, 670 nm); the seventh subsection tagged 'g' show the scattered plot of AOD at 440 nm against number of days; the eighth subsection tagged 'h' show the scattered plot of AOD at 550 nm against number of days; the ninth subsection tagged 'i' show the scattered plot of AOD at 670 nm against number of days; the tenth subsection tagged 'j' show the scattered plot of AOD at 865 nm against number of days; the eleventh subsection tagged 'k' show the 3D image of AOD at 865 nm against AOD at 670 nm; the twelfth subsection tagged 'l' show the 3D image of AOD at 550 nm against AOD at 440 nm; the thirteenth subsection tagged 'm' show the 3D image of AOD at 670 nm against AOD at 550 nm; the fourteenth subsection tagged 'n' show the 3D image of AOD at 865 nm against AOD at 440 nm; the fifteenth subsection tagged 'o' show the 3D image of AOD at 670 nm against AOD at 440 nm; the sixteenth subsection tagged 'p' show the 3D image of AOD at 865 nm against AOD at 670 nm; the seventeenth subsection tagged 'q' show the 2D image of AOD at 440, 550, 670 and 865 nm against the number of days between 2000 and 2013.

In Fig. 4.43, there was high aerosol concentration in some parts on Benin. To understand the extent of pollution, Parakou and Save were chosen as shown in Figs. 5.2 and 5.3 respectively. Figure 5.2a reveals a linear connection between the AOD at 550 with the mean AOD. This may possibly mean that the deviation from the mean is insignificant or very low. Hence, in reality, much assertion cannot be made on the possibility of comparing the AOD at 550 nm and mean AOD to understand the individual AOD expressed in Eq. (1.7). Figure 5.2b show significant scattered distribution. Hence, AOD at 865 may be use to understand the AOD of individual aerosol components expressed in Eq. (1.7). While the AOD at 440 nm against the sum of other band-AOD showed scattered distribution (Fig. 5.2c), the AOD at 550 nm against the sum of other band-AOD showed a linear representation (Fig. 5.2d). Figure 5.2e show the linear relationship between AOD at 670 nm and sum of AOD that represents other AOD band. AOD at 865 nm (like AOD at 440 nm) had a scattered distribution as observed in Fig. 5.2f. The scattered plot for all band is shown in Fig. 5.2g–j. AOD at 865 nm had the most coherent distribution (see color map representation in Fig. 5.2j). AOD at 670 nm (Fig. 5.2i), AOD at 440 nm (Fig. 5.2g) and AOD at 550 nm (Fig. 5.2h) had coherent distribution in descending order.

The interdependency of the AOD of each band is discussed in Fig. 5.2k–p. The interpretation of Fig. 5.2k–p is done from the surface and the shapes within the 3D image. The plot of AOD at 865 nm and AOD at 670 nm (Fig. 5.2k) has high AOD presence between 0.1 and 0.6. The highest and lowest AOD frequency can be found in 0.2–0.3 and 0.8–1.2 respectively. This means the combination of AOD at 865 and AOD at 670 will yield low results when used to estimate aerosol parameters e.g. Angstrom exponent. The plot of AOD at 550 nm and AOD at 440 nm (Fig. 5.2l) have

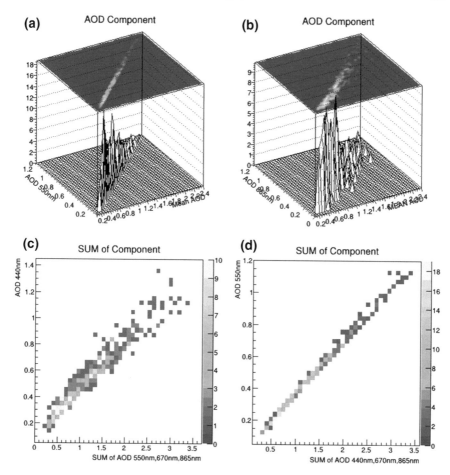

Fig. 5.2 a–d Aerosol inter-relationality I. **e–h** Aerosol daily performance I. **i–l** Aerosol inter-relationality II. **m–p** Aerosol inter-relationality III. **q** Virtual performance of individual AOD

high AOD presence between 0.1 and 0.85. The highest and lowest AOD frequency can be found in 0.2–0.8 and 1.0–1.2 respectively. This means that the Angstrom exponent for each point will be more.

The plot of AOD at 670 nm and AOD at 550 nm (Fig. 5.2m) have high AOD presence between 0.2 and 0.4. The highest and lowest AOD frequency can be found in 0.2–0.5 and 1.0–1.2 respectively.

This means that the Angstrom exponent for each point will be less. Also, Fig. 5.2m is adjudged the most linear and scanty plot. Hence, AOD at 670 and 550 nm has the less dependency on each other. The plot of AOD at 865 nm and AOD at 440 nm (Fig. 5.2n) have high AOD presence between 0.2 and 0.9. The highest and lowest AOD frequency can be found in 0.2–0.9 and 1.0–1.2 respectively. The uniqueness

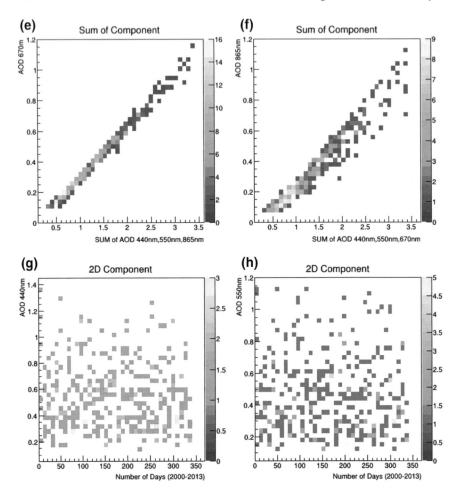

Fig. 5.2 (continued)

of this plot lies in the scattered distribution of its components. Unlike Fig. 5.2m, n has the highest dependency on each other.

The plot of AOD at 670 nm and AOD at 440 nm (Fig. 5.2o) has high AOD presence between 0.2 and 0.7. The highest and lowest AOD frequency can be found in 0.2–0.7 and 1.0–1.2 respectively. Figure 5.2p also shares almost same traits as Fig. 5.2o. The difference between both diagrams is that the scattering of the AOD points in Fig. 5.2p spreads out of its linearity. On the other hand, Fig. 5.2o shows higher dependency on the AOD bands i.e. next to Fig. 5.2n. Figure 5.2q shows the individual performance of the AOD band between 2000 and 2013. Days that had AOD ≥ 1.0 was documented to understand the sequence at which the anthropogenic source influences the AOD's magnitude.

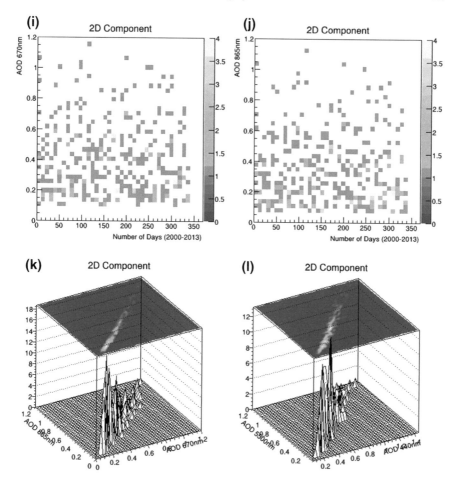

Fig. 5.2 (continued)

The peak of ≥ 1.0 (of AOD) were noted on the following days: 16-March-2000 (day-3), 22-March-2002 (day-50), 9-March-2003 (day-71), 11-March-2004 (day-96), 18-March-2004 (day-97), 5-March-2005 (day-121), 10-March-2006 (day-147), 4-March-2007 (day-175), 15-March-2008 (day-201), 3-March-2010 (250), 21-March-2010 (day-252), 11-Feb-2011 (day-273), 7-April-2011 (day-280), 5-Feb-2012 (day-299), and 17-March-2012 (day-304). It could be concluded that the highest AOD over Parakou occurs every March of the year. No peaks appeared in 2009 and 2013 while two peaks appeared in 2011 within two unusual months (i.e. February and April). It can be inferred that 2009 is unique in the sense that the anthropogenic pollution or aerosol retention might have decreased.

Figure 5.3a–f has same technical concept as Fig. 5.2a–f. The scattered plot for all band is shown in Fig. 5.3g–j differ from Fig. 5.2g–j. Figure 5.3h had the most

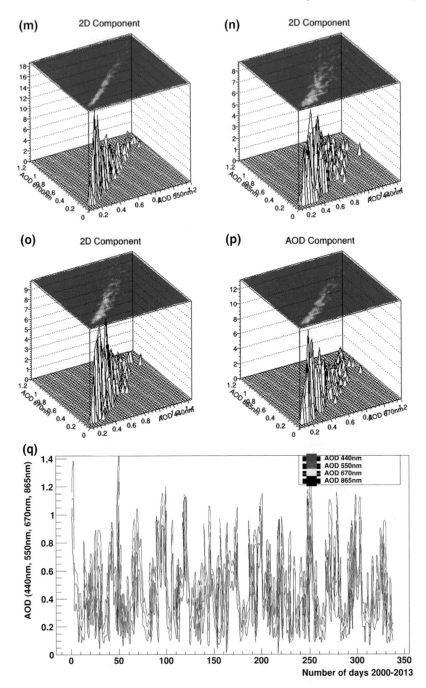

Fig. 5.2 (continued)

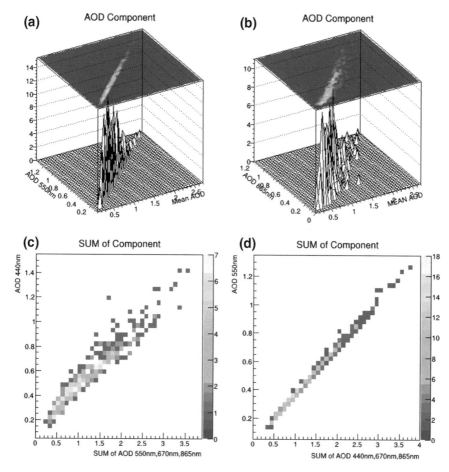

Fig. 5.3 a–d Aerosol inter-relationality I. **e–h** Aerosol daily performance I. **i–l** Aerosol inter-relationality II. **m–p** Aerosol inter-relationality III. **q** Virtual performance of individual AOD

coherent distribution (see color map representation in Fig. 5.3h). AOD at 875 nm (Fig. 5.2j), AOD at 440 nm (Fig. 5.2g) and AOD at 670 nm (Fig. 5.2i) had coherent distribution in descending order. The interdependency of the AOD (Fig. 5.3k–p) of each band has same trend as Fig. 5.2k–p.

The peaks of ≥1.0 (of AOD) were noted on the following days: 7-March-2000 (day-2), 21-Jan-2001 (day-20), 11-April-2001 (day-28), 24-Jan-2002 (43), 22-March-2002 (day-51), 16-March-2003 (day-74), 11-March-2004 (day-96), 29-Nov-2004 (day-107), 17-Feb-2005 (day-109), 5-March-2005 (day-121), 19-Jan-2006 (day-139), 20-Feb-2006 (day-144), 4-March-2007 (day-174), 26-Dec-2008 (day-219), 4-Feb-2011 (day-273), 8-March-2011 (day-275), 19-Dec-2011 (day-287), 20-Jan-2012 (day-293), 17-March-2012 (day-300).

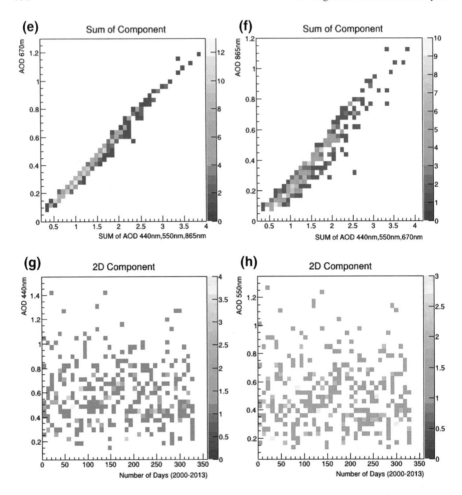

Fig. 5.3 (continued)

It could be concluded that the highest AOD over Save occurs in the month of March over the years. However, there were no peaks in 2009, 2010 and 2013 while two peaks appeared in 2001, 2002, 2004, 2005, 2011 and 2012. It is therefore corroborated by Fig. 5.3q that 2009 had decreased anthropogenic pollution or aerosol retention.

Figures 4.8, 4.15, 4.29 and 4.43 constantly showed the influence of the Sahara Desert at the boundaries of Niger and Chad. It is clear that aside Sahara dust, there are human activities that led to the deposition of sulfate and black carbon over the region. It is in this light we considered close town in Chad (such as Faya and Mao) and Niger (Agadez).

Figure 5.4a–f has same technical concept as Fig. 5.2a–f. The scattered plot for all band is shown in Fig. 5.4g–j. Compared to Figs. 5.2g–j and 5.3g–j, the scattered plot in Fig. 5.4g–j has more data points because its satellite dataset retrieval is not hindered

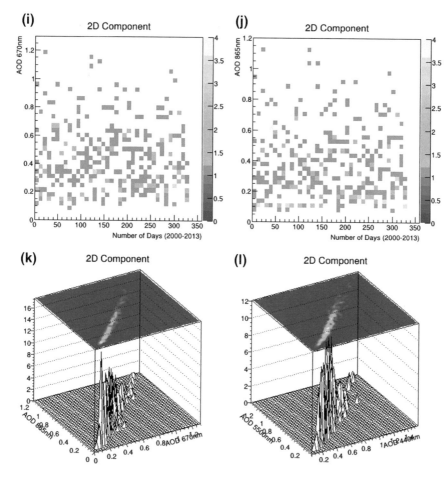

Fig. 5.3 (continued)

by moist (Emetere 2016a, b). Satellite exploration do not retrieve dataset for everyday of the year because of technical issues as orbiting time, moist etc. This occurrence is referred to as 'data loss'. It has been observed that there is a large volume of 'data loss' over West Africa (Emetere et al. 2015a, b, c, d, 2016; Emetere and Akinyemi 2017). Also, there are more 'data loss' in southern-coastline parts of West Africa than in its northern part. Figure 5.4i had the most coherent distribution (AOD at 670 nm). AOD at 550 nm (Fig. 5.4h), AOD at 865 nm (Fig. 5.4j) and AOD at 440 nm (Fig. 5.4g) had coherent distribution in descending order. The interdependency of the AOD (Fig. 5.4k–p) of each band has low coincidence with one another. This means the generation of the dataset for each bands are very dynamic due to the large dispersion source (Sahara Desert).

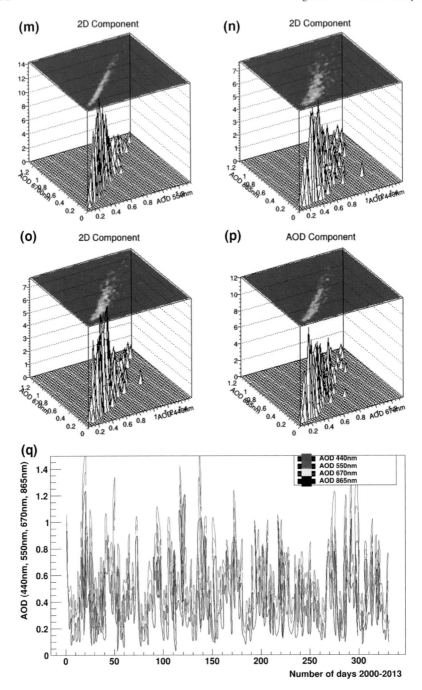

Fig. 5.3 (continued)

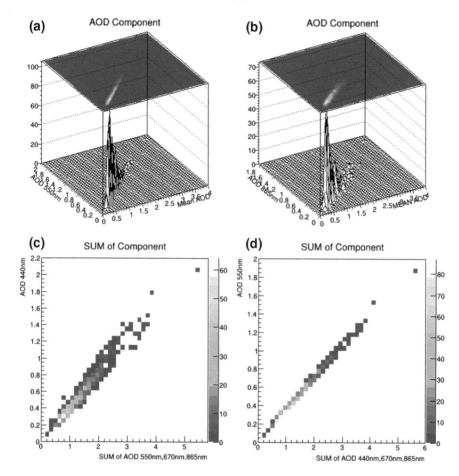

Fig. 5.4 a–d Aerosol inter-relationality I. **e–h** Aerosol daily performance I. **i–l** Aerosol inter-relationality II. **m–p** Aerosol inter-relationality III. **q** Virtual performance of individual AOD

Figure 5.4q has more unique feature i.e. sinusoidal. This result is same for Moa-Chad (Appendix: Fig. B.1) and Agadez-Niger (Appendix: Fig. B.2). The comparative analysis of the three locations (Faya, Mao and Agadez is shown in Table 5.1. The date presented are days when the AOD over the location exceeds 1 i.e. AOD ≥ 1. As discussed earlier, there were lots of data loss, so the number of day presented (for example day-24) refer to the day the satellite retrieved meaningful dataset during the year. The observation drawn from Table 5.1 gives detail on the satellite imagery shown in Figs. 4.8, 4.15, 4.29 and 4.43.

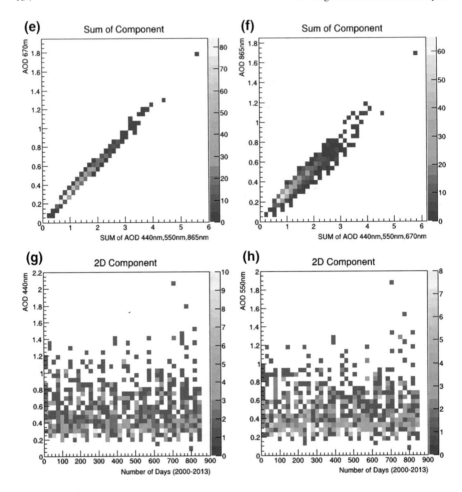

Fig. 5.4 (continued)

The observations are:

i.   Satellite AOD retrieval takes place in the following order i.e. Faya-Chad, Mao-Chad and Agadez-Niger. This may be adduced to the satellite orbiting time over the locations;

ii. Mao-Chad had the most frequency of AOD ≥ 1;

iii. Faya-Chad had more dataset than Mao-Chad and Agadez-Niger;

iv. In 2003, 2005 and 2007, all the location had AOD ≥ 1 in the month of April;

v.   Satellite AOD retrieval takes place in the reversed order i.e. Agadez-Niger, Faya-Chad and Mao-Chad. This may be adduced to the satellite orbiting time over the locations or aerosol layer transport in the atmosphere;

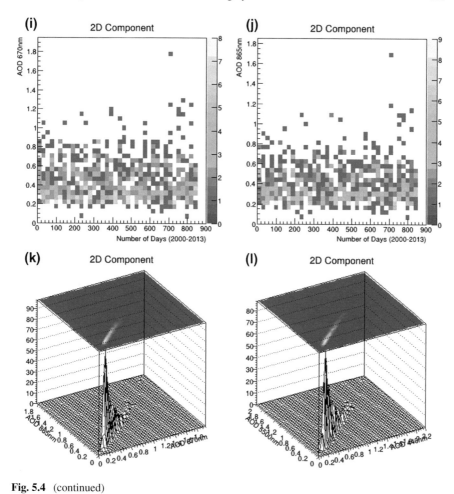

Fig. 5.4 (continued)

vi. The most active month in the three locations is April. The active months are
     listed in descending order i.e. June, August and July. Emetere (2016a, b, 2017a,
     b, c, d) postulated that the months mentioned may be due to aerosol retention
     from October to March.

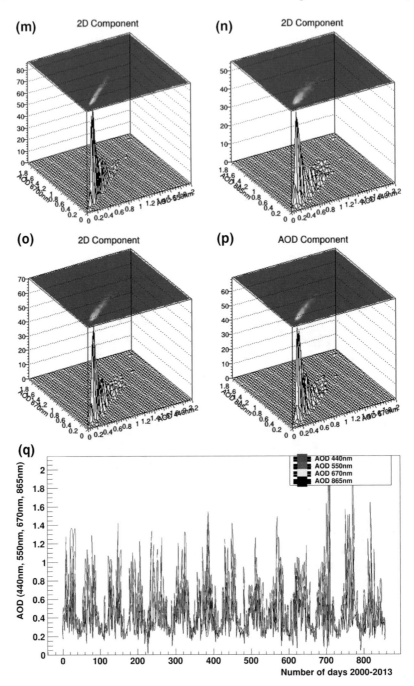

Fig. 5.4 (continued)

5.3   Data Analysis: General Comments on Locations

The interpretation of simulation in environmental science (as shown above) depends on the minor details within its confines and the prevailing situation that may generally lead to a minute or significant change in the simulation. For example, the specific location that were selected in chapter five was based on the information that was retrieved from the satellite images in chapter four. The quality of interpretation is directly proportional to the information that can be retrieved from the available simulations.

One of the cardinal focus of this book is to maximum information that can be retrieved from 'big data'. For example, if a terabyte datasets of images (say >30,000) is to be analyzed, it is advisable to use the following rules:

i.   Do the statistical analysis of the images (see Figs. 4.53–4.55). This process engenders more information on the general trends of the curves. The trend may either satisfy a given condition or known event. For example, aerosols optical depth is expected to be high from October to April in the tropical region of west Africa. Also, the sudden increase in aerosol retention from June to August (Fig. 5.1) gives indication of the existence of an unknown event in the geographical area. Hence, seeking adequate interpretation for the trends may allow you narrow to specific points in the simulation.

ii. Do the multi-dimensional analysis of each images located on the hotspot. For example, Figs. 4.1–4.52 will show the significant differences or similarities between chosen images. This section can only be successful if adequate attention is given to the program or macro that is used for the simulation. It is advisable to numerically analyze the images so you can quantify the 'big data' inform of computational or mathematical model. For example, Figs. 5.5 and 5.6 show the intensity and deviations obtained from 3600 images. Many more processes that can lead to more quality information can be extracted. This information can be categorized as derived and basic. Figure 5.5 is a basic information that can be obtained directly from all the images while Fig. 5.6 is the product of a basic parameter (i.e. derived information). Figure 5.6 may be obtained from the characterization of the pixel of each images or the intensity of the images. The author, believes that the deviation via pixel characterization is the best way of spotting the deviation between images. An example of the first fifteen pixel of three grayscale images is shown Figs. 5.7.

iii. The validation of the results obtained from the re-processed satellite images was achieved by considering the analysis of the '.txt' file for each location. Aside validating assertions, it can be used to obtain detailed information of the hotspot. For example, Table 5.1 show that more anthropogenic activity occurs in Chad (Mao) within the three locations that is close to the Sahara Desert. When designing a project in environmental science (may be using 'big data'), it is advisable to provide means of validation. Also, it is necessary to design the program or macro to generate formidable simulations that will ultimate improve the quality of the project.

Table 5.1 Interdependency of cities close to the Sahara at AOD $\geq$ 1

Year	Chad (Faya)	Chad (Mao)	Niger (Agadez)
2000	20-April-2000 (day-9), 1-August-2000 (day-25)	6-August-2000 (day-20), 25-September-2000 (day-28)	Nil
2001	21-April-2001 (day-68)	19-April-2001 (day-56), 30-May-2001 (day-60)	2-June-2001 (day-46)
2002	26-April-2002 (day-129), 23-August-2002 (day-148)	21-March-2002 (day-103), 1-May-2002 (day-107), 13-July-2002 (day-118), 6-September-2002 (day-125), 22-September-2002 (day-126)	12-June-2002 (day-87)
2003	20-April-2003 (day-189), 10-August-2003 (day-208)	4-February-2003 (day-148), 11-April-2003 (day-159), 13-May-2003 (day-164), 27-May-2003 (day-166)	28-April-2003 (day-119)
2004	5-March-2004 (day-237), 31-May-2004 (day-253), 27-July-2004 (day-264)	16-February-2004 (day-201), 13-May-2004 (day-213), 16-July-2004 (day 221)	1-June-2004 (day-157)
2005	25-April-2005 (day-309), 8-August-2005 (day-328)	23-April-2005 (day-260)	26-April-2005 (day-193)
2006	28-April-2006 (day-372), 24-July-2006 (day-388)	9-March-2006 (day-300), 3-May-2006 (day-309)	16-June-2006 (day-239)
2007	6-April-2007 (day-432), 27-July-2007 (day-452)	13-April-2007 (day-362), 15-May-2007 (day-367), 2-July-2007 (day-376), 6-October-2007 (day-387)	7-April-2007 (day-268)
2008	6-July-2008 (day-511)	26-May-2008 (day-416), 29-September-2008 (day-432)	18-April-2008 (day-306)
2009	23-June-2009 (day-569)	24-March-2009 (day-456), 30-July-2009 (day-471), 8-August-2009 (day-472)	17-July-2009 (day-355)
2010	23-April-2010 (day-624), 14-September-2010 (day-647)	20-March-2010 (day-508)	8-April-2010 (day-383)
2011	7-August-2011 (day-707)	15-April-2011 (day-561), 20-June-2011 (day-570)	26-March-2011 (day-422)
2012	6-June-2012 (day-762), 9-August-2012 (day-773)	16-March-2012 (day-608), 18-March-2012 (day-609)	7-June-2012 (day-472)
2013	8-May-2013 (day-820)	7-June-2013 (day-663), 25-July-2013 (day-668)	3-June-2013 (day-514)

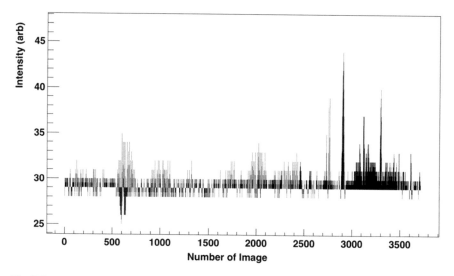

Fig. 5.5   The intensity obtained from 3600 images

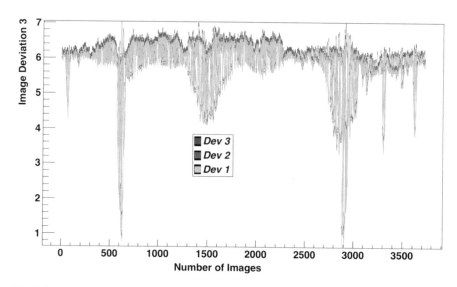

Fig. 5.6   The deviations obtained from 3600 images

(0,0) 38205 38205 38205	(0,0) 22333 22333 22333	(0,0) 6205 6205 6205
(0,1) 25019 25019 25019	(0,1) 45147 45147 45147	(0,1) 61435 61435 61435
(0,2) 3200 3200 3200	(0,2) 2816 2816 2816	(0,2) 2560 2560 2560
(0,3) 3200 3200 3200	(0,3) 3584 3584 3584	(0,3) 3840 3840 3840
(0,4) 3712 3712 3712	(0,4) 3712 3712 3712	(0,4) 4864 4864 4864
(0,5) 3968 3968 3968	(0,5) 5504 5504 5504	(0,5) 4224 4224 4224
(0,6) 4096 4096 4096|	(0,6) 4224 4224 4224	(0,6) 3712 3712 3712
(0,7) 3200 3200 3200	(0,7) 3712 3712 3712	(0,7) 3584 3584 3584
(0,8) 3840 3840 3840	(0,8) 2944 2944 2944	(0,8) 4096 4096 4096
(0,9) 4352 4352 4352	(0,9) 4096 4096 4096	(0,9) 3328 3328 3328
(0,10) 3968 3968 3968	(0,10) 4352 4352 4352	(0,10) 4480 4480 4480
(0,11) 3840 3840 3840	(0,11) 3200 3200 3200	(0,11) 3328 3328 3328
(0,12) 4736 4736 4736	(0,12) 3968 3968 3968	(0,12) 4224 4224 4224
(0,13) 3712 3712 3712	(0,13) 4224 4224 4224	(0,13) 3584 3584 3584
(0,14) 3584 3584 3584	(0,14) 2944 2944 2944	(0,14) 3840 3840 3840
Image 1	Image 2	Image 3

Fig. 5.7 Characterization of image pixel

5.4 Designing the Code to Analyze Big Data

The modalities in this section is same as the Sect. 4.3. In this section, two types of macros shall be described. The format of big data discussed in this section is '.txt', '.dat' etc. The data may come in a raw form i.e. many unwanted information within rows or columns. In actual fact, most datasets obtained from primary sources like satellite companies, communication companies etc. are in a raw format which may be irritating to edit. Editing a raw dataset of one terabyte and above is tedious and fool-hardy. The macro below shows how the author edited raw dataset within the code.

5.4.1 *Macro One*

```
void Macro1(){
        Float_t Channel,ChannelContent,z,Channel1,ChannelContent1,z1;
        Int_t ncols,ncols1;
        Int_t nlines = 0,nlines1 = 0;
        char line[3000];  // As large as the lines that you are reading in

        // Read from .dat file.

        ifstream datFile;
        ifstream datFile1;
        datFile.open("Space_ValveClosed_mca1.dat");
        datFile1.open("Space_ValveOpen_mca1.dat");
        // skip the first two lines of the dataset


        datFile.getline(line,128);
        datFile.getline(line,128);
        datFile1.getline(line,128);
        datFile1.getline(line,128);

        TFile*f = new TFile("Space_ValveClosed_mca1","RECREATE");
```

```
TFile *f1 = new TFile("Space_ValveOpen_mca1","RECREATE");
TH1F *h = new TH1F("h","Channel distribution",100,-4,4);
TH1F *h1 = new TH1F("h1","Channel distribution",100,-4,4);
TNtuple      *ntuple      =      new      TNtuple("ntuple","data      from
MCA","Channel:ChannelContent");
TNtuple      *ntuple1     =      new      TNtuple("ntuple1","data     from
MCA","Channel1:ChannelContent1");
while (1) {
    datFile >> Channel >> ChannelContent;
    if (!datFile.good()) break;
    if (nlines < 5) printf("x=%8f, y=%8f\n",Channel,ChannelContent);
    h->Fill(Channel);
    ntuple->Fill(Channel,ChannelContent);
    nlines++;
}
printf(" found %d points\n",nlines);
 datFile.close();

  printf(line, "%f\n",z);

while (2) {
    datFile1 >> Channel1 >> ChannelContent1;
    if (!datFile1.good()) break;
    if (nlines1 < 5) printf("x=%8f, y=%8f\n",Channel1,ChannelContent1);
    h->Fill(Channel1);
    ntuple->Fill(Channel1,ChannelContent1);
    nlines1++;
}
printf(" found %d points\n",nlines1);
 datFile1.close();

  printf(line, "%f\n",z);

    TCanvas *MyCanvas = new TCanvas("canv", "General Plots",800,600);
```

```
                TCanvas *MyCanvas1 = new TCanvas("canv1", "General Plots",800,600);

                ntuple->Draw("ChannelContent:Channel");
                ntuple1->Draw("ChannelContent1:Channel1");

                TCanvas *MyCan = new TCanvas("ca", "General Plots",800,600);
                TCanvas *MyCan1 = new TCanvas("ca1", "General Plots",800,600);
                //MyCanvas->Divide(2,1);
                //MyCanvas->cd(1);
                ntuple1->Draw("ChannelContent/2075.007998:Channel");
                TH2F *htemp = (TH2F*)gPad->GetPrimitive("htemp");
                htemp->GetXaxis()->SetTitle("Channel");
                htemp->GetYaxis()->SetTitle("Channel Content per time [count/sec]");
                htemp->SetFillColor(42);
               htemp->SetMarkerColor(3);
          htemp->SetMarkerStyle(3);
                    htemp->SetTitle("MCA Plots");
                      f->Write();
    }
```

5.4.2   Macro Two

The second type of dataset is the processed dataset. This kind of data can be obtained from environmental monitoring centers etc. Irrelevant information is littered within rows and column. Hence, the way of writing the codes differs as shown below.

```
void Macro2(const char *dirname="testdata.txt", const char *ext=".dat"){
     TString dir = gSystem->UnixPathName(gInterpreter->GetCurrentMacroName());
     dir.ReplaceAll("Macro2.C","");
     dir.ReplaceAll("/./","/");
     TFile *f = new TFile("dirname.root","RECREATE"); //create file data.root
     TTree *tree = new TTree("tree","data from ascii file");
     Double_t                    nlines                =                    tree-
>ReadFile(Form(dirname,dir.Data()),"pa:pb:pc:pd:pe:pf:pg:ph:pi:pj:pk;pl:pm:pn:po:pp:pq:pr:
```

```
ps:pt:pu:pv:pw:px:py,pz");//create tree with
        gROOT->SetStyle("Plain");
        gStyle->SetOptStat(1111);
        gStyle->SetOptFit(1111);
        TCanvas *c = new TCanvas("c", "General Plots1",800,600);
        c->Divide(2,2);
        c->SetFillColor(5);
        c->SetFrameFillColor(10);
        TMultiGraph * mg = new TMultiGraph("mg","mg");
      //make graphs
        c->cd(1);
      tree->Draw("pb:pa");
        TGraph *gr = new TGraph(tree->GetSelectedRows(),tree->GetV2(), tree->GetV1());
        gr->SetName("myGraph");
        /*TH2F *htemp = (TH2F*)gPad->GetPrimitive("htemp");
        htemp->GetXaxis()->SetTitle("Channel");
        htemp->GetYaxis()->SetTitle("Channel Content per time");
        htemp->SetFillColor(42);
        htemp->SetMarkerColor(3);
    htemp->SetMarkerStyle(7);
            htemp->SetTitle("MCA Plots");*/
        gr->Draw();

      c->cd(2);
      tree->Draw("pc:pa");
      TGraph *gr2 = new TGraph(tree->GetSelectedRows(),tree->GetV2(), tree->GetV1());
        gr2->SetName("myGraph");
      gr2->Draw();
      c->cd(3);
      tree->Draw("pd:pa");
        TGraph *gr3 = new TGraph(tree->GetSelectedRows(),tree->GetV2(), tree->GetV1());
      gr3->SetName("myGraph");
      gr3->Draw();
```

```
      c->cd(4);
      tree->Draw("pe:pa");
      TGraph *gr4 = new TGraph(tree->GetSelectedRows(),tree->GetV2(), tree->GetV1());
      gr4->SetName("myGraph");
      gr4->Draw();
      c->Update();
   TImage *img = TImage::Create();
    img->FromPad(c);
    img->WriteImage("canvas1.png");


      TCanvas *c1 = new TCanvas("c1", "General Plots2",800,600);
      c1->Divide(2,2);
      c1->cd(1);
      tree->Draw("pf:pa");
      TGraph *jr = new TGraph(tree->GetSelectedRows(),tree->GetV2(), tree->GetV1());
      jr->SetName("myGraph");
      jr->Draw();

      c1->cd(2);
      tree->Draw("pg:pa");
      TGraph *jr2 = new TGraph(tree->GetSelectedRows(),tree->GetV2(), tree->GetV1());
      jr2->SetName("myGraph");
      jr2->Draw();

      c1->cd(3);
      tree->Draw("ph:pa");
      TGraph *jr3 = new TGraph(tree->GetSelectedRows(),tree->GetV2(), tree->GetV1());
      jr3->SetName("myGraph");
      jr3->Draw();

      c1->cd(4);
      tree->Draw("pi:pa");
      TGraph *jr4 = new TGraph(tree->GetSelectedRows(),tree->GetV2(), tree->GetV1());
      jr4->SetName("myGraph");
```

```
    jr4->Draw();
  TImage *imj = TImage::Create();
   imj->FromPad(c1);
   imj->WriteImage("canvas2.png");


    TCanvas *c2 = new TCanvas("c2", "General Plots3",800,600);
    c2->Divide(2,2);
    c2->cd(1);
    tree->Draw("pj:pa");
    TGraph *pr = new TGraph(tree->GetSelectedRows(),tree->GetV2(), tree->GetV1());
    pr->SetName("myGraph");
    pr->Draw();


    c2->cd(2);
    tree->Draw("pk:pa");
    TGraph *pr2 = new TGraph(tree->GetSelectedRows(),tree->GetV2(), tree->GetV1());
    pr2->SetName("myGraph");
    pr2->Draw();


    c2->cd(3);
    tree->Draw("pl:pa");
    TGraph *pr3 = new TGraph(tree->GetSelectedRows(),tree->GetV2(), tree->GetV1());
    pr3->SetName("myGraph");
    pr3->Draw();


    c2->cd(4);
    tree->Draw("pm:pa");
    TGraph *pr4 = new TGraph(tree->GetSelectedRows(),tree->GetV2(), tree->GetV1());
    pr4->SetName("myGraph");
    pr4->Draw();
  TImage *imp = TImage::Create();
   imp->FromPad(c2);
   imp->WriteImage("canvas3.png");
```

```
TCanvas *c3 = new TCanvas("c3", "General Plots4",800,600);
c3->Divide(2,2);
c3->cd(1);
tree->Draw("pn:pa");
TGraph *kr = new TGraph(tree->GetSelectedRows(),tree->GetV2(), tree->GetV1());
kr->SetName("myGraph");
kr->Draw();

c3->cd(2);
tree->Draw("po:pa");
TGraph *kr2 = new TGraph(tree->GetSelectedRows(),tree->GetV2(), tree->GetV1());
kr2->SetName("myGraph");
kr2->Draw();

c3->cd(3);
tree->Draw("pp:pa");
TGraph *kr3 = new TGraph(tree->GetSelectedRows(),tree->GetV2(), tree->GetV1());
kr3->SetName("myGraph");
kr3->Draw();

c3->cd(4);
tree->Draw("pq:pa");
TGraph *kr4 = new TGraph(tree->GetSelectedRows(),tree->GetV2(), tree->GetV1());
kr4->SetName("myGraph");
kr4->Draw();
TImage *imk = TImage::Create();
imk->FromPad(c3);
imk->WriteImage("canvas4.png");

TCanvas *c4 = new TCanvas("c4", "General Plots5",800,600);
c4->Divide(2,2);
c4->cd(1);
tree->Draw("pr:pa");
TGraph *fr = new TGraph(tree->GetSelectedRows(),tree->GetV2(), tree->GetV1());
```

```
fr->SetName("myGraph");
fr->Draw();


c4->cd(2);
tree->Draw("ps:pa");
TGraph *fr2 = new TGraph(tree->GetSelectedRows(),tree->GetV2(), tree->GetV1());
fr2->SetName("myGraph");
fr2->Draw();


c4->cd(3);
tree->Draw("pt:pa");
TGraph *fr3 = new TGraph(tree->GetSelectedRows(),tree->GetV2(), tree->GetV1());
fr3->SetName("myGraph");
fr3->Draw();


c4->cd(4);
tree->Draw("pu:pa");
TGraph *fr4 = new TGraph(tree->GetSelectedRows(),tree->GetV2(), tree->GetV1());
fr4->SetName("myGraph");
fr4->Draw();
TImage *imf = TImage::Create();
imf->FromPad(c4);
imf->WriteImage("canvas5.png");


TCanvas *c5 = new TCanvas("c5", "General Plots6",800,600);
c5->Divide(3,2);
c5->cd(1);
tree->Draw("pv:pa");
TGraph *dr = new TGraph(tree->GetSelectedRows(),tree->GetV2(), tree->GetV1());
dr->SetName("myGraph");
dr->Draw();


c5->cd(2);
tree->Draw("pw:pa");
```

```
    TGraph *dr2 = new TGraph(tree->GetSelectedRows(),tree->GetV2(), tree->GetV1());
    dr2->SetName("myGraph");
    dr2->Draw();


    c5->cd(3);
    tree->Draw("px:pa");
    TGraph *dr3 = new TGraph(tree->GetSelectedRows(),tree->GetV2(), tree->GetV1());
    dr3->SetName("myGraph");
    dr3->Draw();


    c5->cd(4);
    tree->Draw("py:pa");
    TGraph *dr4 = new TGraph(tree->GetSelectedRows(),tree->GetV2(), tree->GetV1());
    dr4->SetName("myGraph");
    dr4->Draw();


    c5->cd(5);
    tree->Draw("pz:pa");
    TGraph *dr5 = new TGraph(tree->GetSelectedRows(),tree->GetV2(), tree->GetV1());
    dr5->SetName("myGraph");
    dr5->Draw();
  TImage *imd = TImage::Create();
   imd->FromPad(c4);
   imd->WriteImage("canvas6.png");

}
```

References

Borne, K. (2014). *Top 10 big data challenges—A serious look at 10 big data V's*. https://mapr.com/blog/top-10-big-data-challenges-serious-look-10-big-data-vs/. Accessed January 31, 2018.

Cleverism. (2018). *What is big data?* https://www.cleverism.com/brief-history-big-data/. Accessed January 31, 2018.

Dataversity. (2018). Big Data Trends for 2018, https://www.dataversity.net/big-data-trends-2018/. Accessed November 12, 2018.

Emetere, M. E. (2016a). Statistical examination of the aerosols loading over Mubi-Nigeria: The satellite observation analysis. *Geographica Panonica, 20*(1), 42–50.

Emetere, M. E. (2016b). *Numerical modelling of West Africa regional scale aerosol dispersion*. Thesis submitted to Covenant University.

Emetere, M. E. (2017a). Investigations on aerosols transport over micro- and macro-scale settings of West Africa. *Environmental Engineering Research, 22*(1), 75–86.

Emetere, M. E. (2017b). Lightning as a source of electricity: Atmospheric modeling of electromagnetic fields. *International Journal of Technology, 8,* 508–518.

Emetere, M. E. (2017c). Impacts of recirculation event on aerosol dispersion and rainfall patterns in parts of Nigeria. *Global Nest Journal, 19*(2), 344–352.

Emetere, M. E. (2017d). Monitoring the 3-year thermal signatures of the Calbuco pre-volcano eruption event. *Arabian Journal of Geoscience, 10,* 94. https://doi.org/10.1007/s12517-017-2861-z.

Emetere, M. E., & Akinyemi, M. L. (2017). Documentation of atmospheric constants over Niamey, Niger: A theoretical aid for measuring instruments. *Meteorological Applications, 24*(2), 260–267.

Emetere, M. E., Akinyemi, M. L., & Akinojo, O. (2015a). A novel technique for estimating aerosol optical thickness trends using meteorological parameters. *2015 PIAMSEE: AIP Conference Proceedings, 1705*(1), 020037.

Emetere, M. E., Akinyemi, M. L., & Uno, U. E. (2015b). Computational analysis of aerosol dispersion trends from cement factory. In *IEEE Proceedings 2015 International Conference on Space Science & Communication* (pp. 288–291).

Emetere, M. E., Akinyemi, M. L., & Akinojo, O. (2015c). Parametric retrieval model for estimating aerosol size distribution via the AERONET, LAGOS station. *Environmental Pollution, 207*(C), 381–390.

Emetere, M. E., Akinyemi, M. L., & Akin-Ojo, O. (2015d). Aerosol optical depth pollution in selected areas trends over different regions of Nigeria: Thirteen years analysis. *Modern Applied Science, 9*(9), 267–279.

Emetere, M. E., Akinyemi, M. L., & Edeghe, E. B. (2016). A simple technique for sustaining solar energy production in active convective coastal regions. *International Journal of Photoenergy, 2016*(3567502), 1–11. https://doi.org/10.1155/2016/3567502.

Foote, K. D. (2017). *A brief history of big data.* http://www.dataversity.net/brief-history-big-data/. Accessed January 30, 2018.

Qubole, (2008). The Future of Big Data and Machine Learning Is Clear: It's All on the Cloud, https://www.qubole.com/blog/the-future-of-big-data-and-machine-learning-is-clear-its-all-on-the-cloud/. Accessed November 12, 2018.

Stephenson, D. (2013). 7 big data techniques that create business value. https://www.firmex.com/thedealroom/7-big-data-techniques-that-create-business-value/. Accessed January 31, 2018.

Chapter 6
Conclusion

In Chap. 1, environmental modelling was explain as an efficient working system that describes in totality the whole working principle of an environmental event. It was emphasized that environmental modelling is one of the largest researched area in academics. The prospects of environment studies have currently expanded beyond the perimeter of science. From the nitty-gritty of the types of environmental modelling, it is easier to accommodate more professions than ever imagined. This means that its application in modern times will be enormous. A typical example was narrowed to the most unpredictable aspect of environmental modeling-atmospheric physics. The research focused on atmospheric aerosols by first considering the known aerosols models. The mathematical dynamics as well as the meteorological implication was illustrated in clear terms.

In Chap. 2, the different computational techniques were discussed. This technique transforms the computer system into a 'wonder-machine' that may be better than the modeler imagination. The interdisciplinary dependency on computer is growing into a big hub of knowledge where research output is increased with less interest in experiments or field work. Big organization now seek ways of optimization by compiling their operation in form of computer application. The acceptability of a computer application depends on its wide usability. Most good computer applications are not accepted not because it is not good but because the inventor may not have the funds to pull through to large audience.

The components of the computer as it relates to its individual functionality was discussed. Based on the above, a review of available environmental software was discussed. Most environmental software is licensed for commercial patronage. However, the open source packages are free to use. However, the disadvantage of the open source application or library is that the user must be versed in the use of one or more computer language i.e. C++, C, C#, Python, Java etc. In this book, the European Organization for Nuclear Research (CERN) software known as 'Root' was used in this book. The Root function-ability was discussed with the view for readers to understand why more science disciplines are attracted to it. An appreciable scope of the open-source library was discussed. The author adopted the OpenCV library. The

© Springer Nature Switzerland AG 2019
M. E. Emetere, *Environmental Modeling Using Satellite Imaging
and Dataset Re-processing*, Studies in Big Data 54,
https://doi.org/10.1007/978-3-030-13405-1_6

scope of the OpenCV was discussed to also show why more science and engineering professional adopt this library for their work.

In Chap. 3, the focus was to discuss on a typical environmental model. Outdoor air pollution was chosen as case study. The theoretical dynamics of air pollution was discussed. In order to understand the complexity of the air pollution, many mathematical, statistical and computational models have been propounded by scientist. The main challenge of outdoor pollution is its open-system i.e. pollutants introduced into the system may deplete via some unexplainable event. This has introduced large errors in the formulation and execution of its model. The outdoor pollution dispersion was holistically reviewed with the understanding that the inadequate representation of the mathematics of the problem has a way of compromising any computer application that was invented from its principles. Lastly, the documentation of the published solutions of large-scale air pollution were shown. The shortcoming of each analytical models were discussed.

The concept of image re-processing of satellite image was introduced by discussing properties of an image pixel, resolution etc. The quality of an image is directly proportional to the quality of the camera. Hence, few satellite camera was discussed. The many types of image format were mentioned. West Africa was considered as a case study in this chapter. The properties of the satellite image produced were shown in a table. The images were re-processed to enhance the information that can be extracted from the image. The advantage of the processed image is unprecedented compared to the raw satellite image. It was agreed that the new information obtained from the re-processed images strengthen the mode of interpretation for each image. New concepts for interpreting re-processed images was propounded and applied. It was observed that the hypothesis holds for over 93% of the cases considered. More numerical interpretation of images was considered. The concept of statistical analysis of large volume of images was illustrated. Lastly an illustration on how to plan a computational code was discussed. It was concluded that a well-planned code yields better outcome. More so, it is very advantageous to apply the code to large dataset. A typical example of a program and macro were given for readers' appreciation of planning a code to analyze big data. An example of computational and mathematical model was shown. The sensitivity of the model was validated.

The concept of big data was examined by considering thirteen years' dataset for one hundred and twenty-five locations across west Africa. The definition of big data in relation to environmental studies was defined. The peculiarity of the dataset was discussed in detail. The section is used to validate known discoveries or observations in Chap. 4. New concept or deeper information were shown when analyzing ASCII dataset. Salient observations were made. There was a typical illustration on how to plan a project using the highlights in this book. Emphasis was working on a large dataset. The two ways of processing precise and clumsy dataset was discussed. Two macros were illustrated for better understanding. It was concluded that the quality of simulation of ASCII datasets depends largely on the uniqueness of the macro or program.

A new computational design for analyzing satellite image and ASCII dataset has been propounded to assist the technical output of workers, scientists, assistants and interns—working in satellite company, university, research centers, communication companies, engineering companies and research institutes. The illustrations, study articulation and results interpretation are consciously designed for readers' appreciation of the beauty of the involvement of computational techniques in environmental modelling.

Appendix A

See Figs. A.1, A.2, A.3, A.4, A.5, A.6, A.7, A.8, A.9, A.10, A.11, A.12, A.13, A.14, A.15, A.16, A.17, A.18, A.19, A.20, A.21, A.22, A.23, A.24, A.25, A.26, A.27, A.28, A.29, A.30, A.31, A.32, A.33, A.34, A.35, A.36, A.37, A.38, A.39, A.40, A.41, A.42, A.43, A.44, A.45, A.46, A.47, A.48, A.49, A.50, A.51, A.52, A.53, A.54, A.55, A.56, A.57, A.58, A.59, A.60, A.61, A.62, A.63, A.64, A.65, A.66, A.67, A.68, A.69, A.70, A.71, A.72, A.73, A.74, A.75, A.76, A.77, A.78, A.79, A.80, A.81 and A.82.

© Springer Nature Switzerland AG 2019   175
M. E. Emetere, *Environmental Modeling Using Satellite Imaging
and Dataset Re-processing*, Studies in Big Data 54,
https://doi.org/10.1007/978-3-030-13405-1

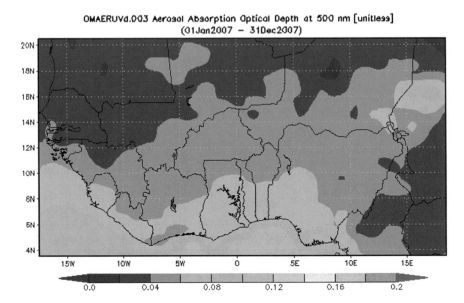

Fig. A.1 Satellite image AAOD at 500 nm, Jan. to Dec., 2007

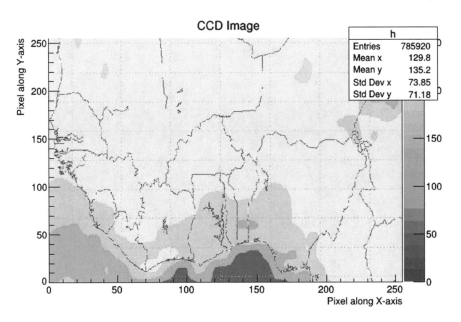

Fig. A.2 x and y pixel redefinition of the satellite image

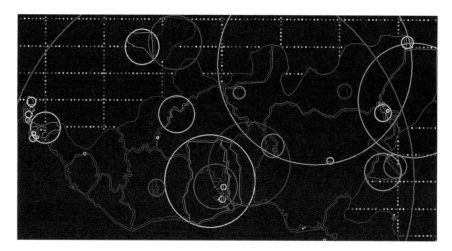

Fig. A.3 Contour detection of satellite image

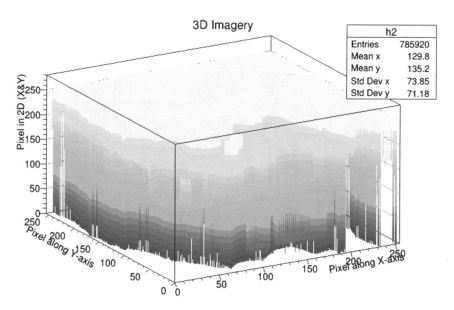

Fig. A.4 3D setting of satellite image

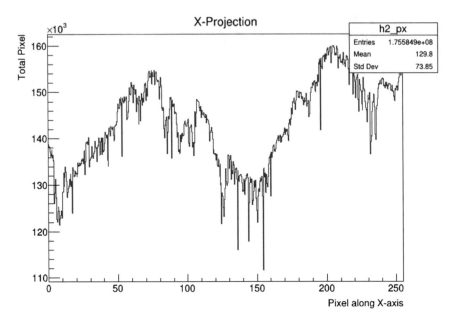

Fig. A.5 X-projection of satellite image

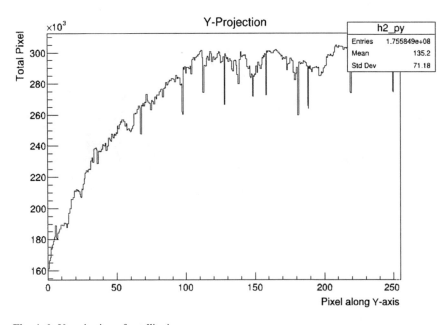

Fig. A.6 Y-projection of satellite image

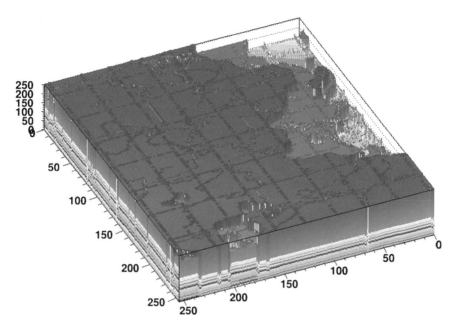

Fig. A.7 The spectrum analysis of the satellite image

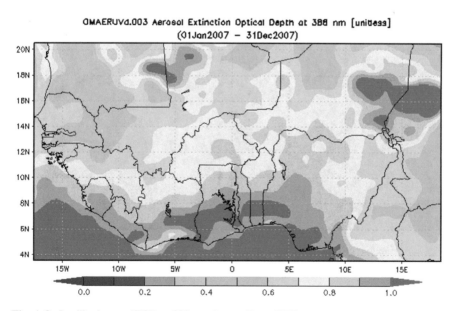

Fig. A.8 Satellite image AEOD at 388 nm, Jan. to Dec., 2007

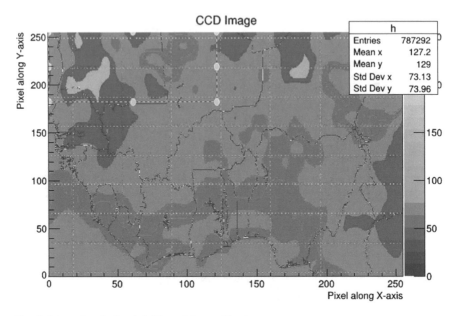

Fig. A.9 x and y pixel redefinition of the satellite image

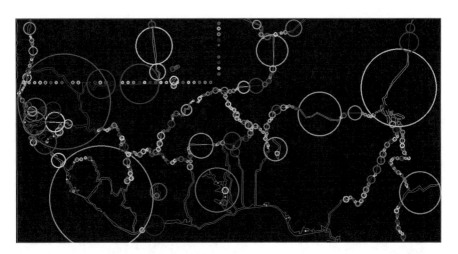

Fig. A.10 Contour detection of satellite image

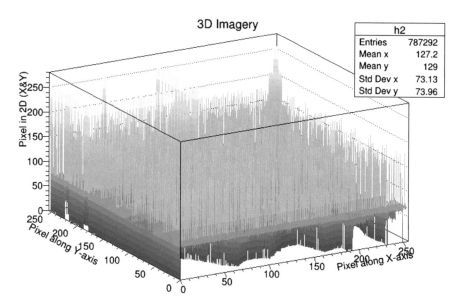

Fig. A.11 3D setting of satellite image

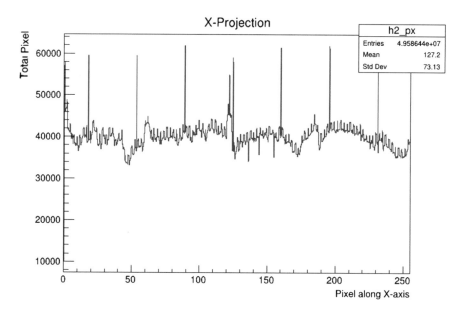

Fig. A.12 X-projection of satellite image

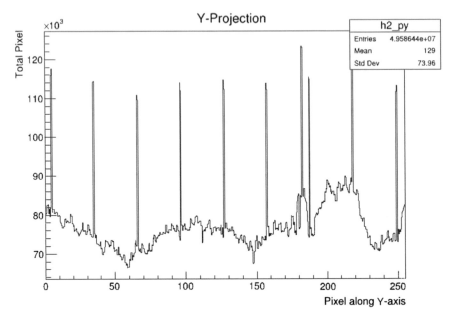

Fig. A.13 Y-projection of satellite image

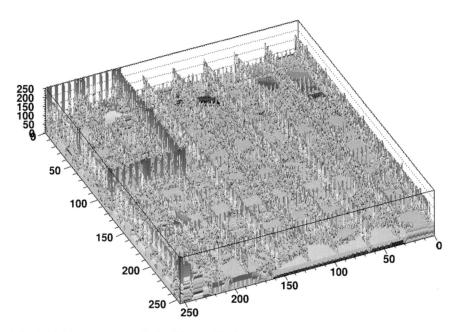

Fig. A.14 The spectrum analysis of the satellite image

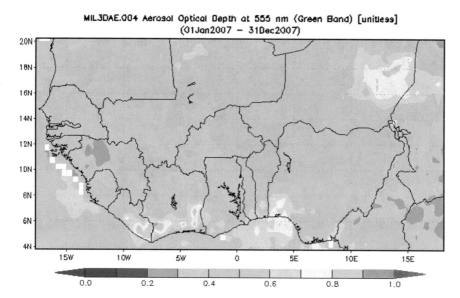

Fig. A.15 Satellite image AOD at 555 nm, Jan. to Dec., 2007

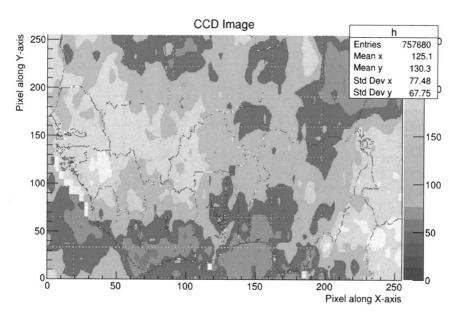

Fig. A.16 x and y pixel redefinition of the satellite image

Fig. A.17 Contour detection of satellite image

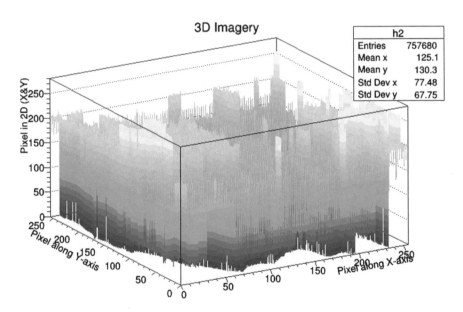

Fig. A.18 3D setting of satellite image

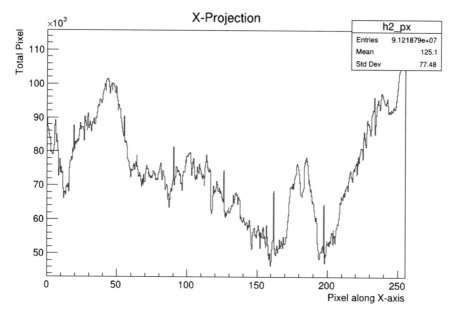

Fig. A.19 X-projection of satellite image

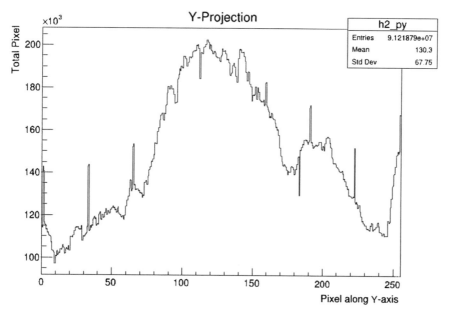

Fig. A.20 Y-projection of satellite image

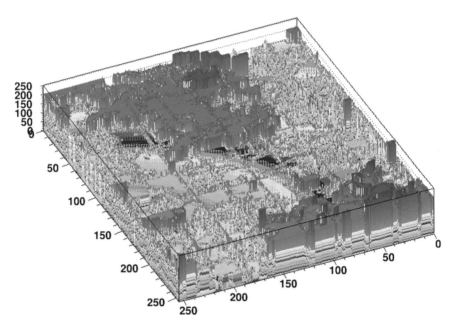

Fig. A.21 The spectrum analysis of the satellite image

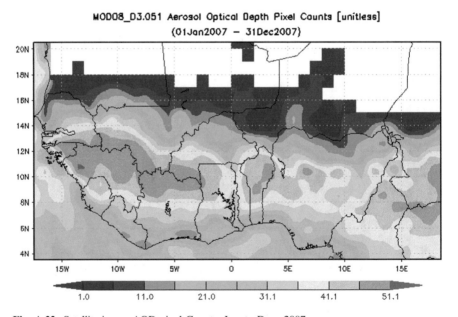

Fig. A.22 Satellite image AOD pixel Counts, Jan. to Dec., 2007

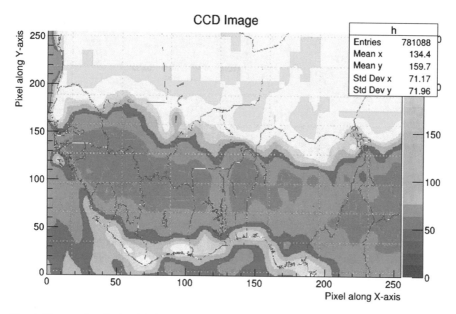

Fig. A.23 x and y pixel redefinition of the satellite image

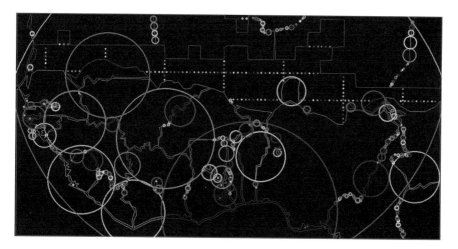

Fig. A.24 Contour detection of satellite image

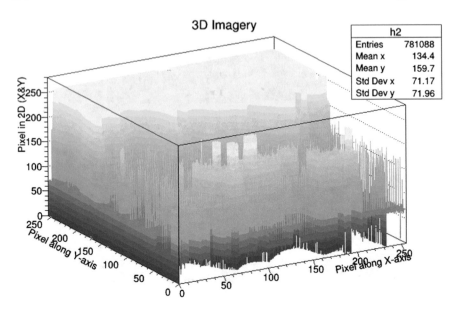

Fig. A.25 3D setting of satellite image

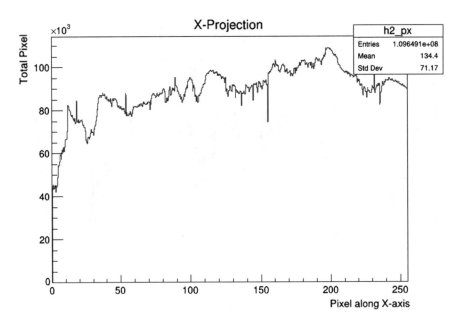

Fig. A.26 X-projection of satellite image

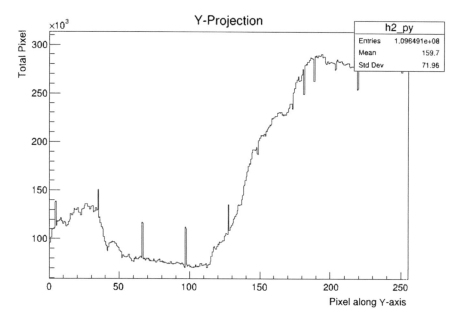

Fig. A.27 Y-projection of satellite image

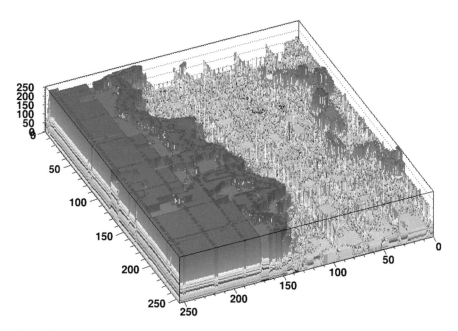

Fig. A.28 The spectrum analysis of the satellite image

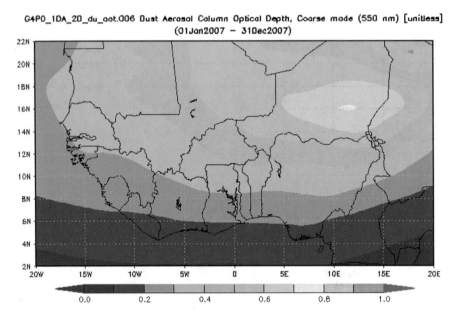

Fig. A.29 Satellite image AOD, Coarse mode, Jan. to Dec., 2007

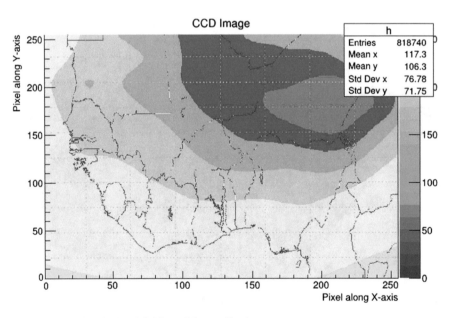

Fig. A.30 x and y pixel redefinition of the satellite image

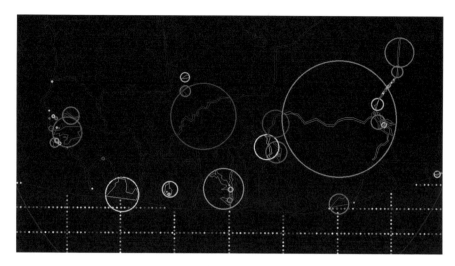

Fig. A.31 Contour detection of satellite image

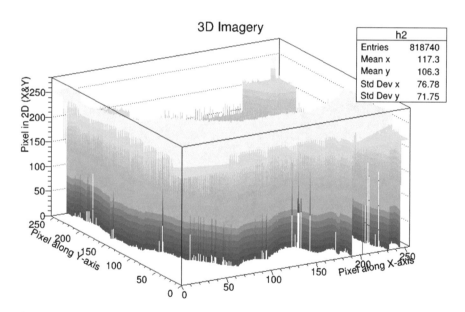

Fig. A.32 3D setting of satellite image

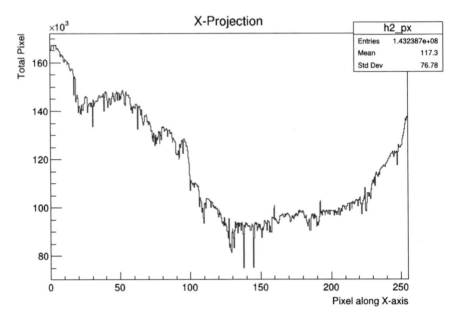

Fig. A.33 X-projection of satellite image

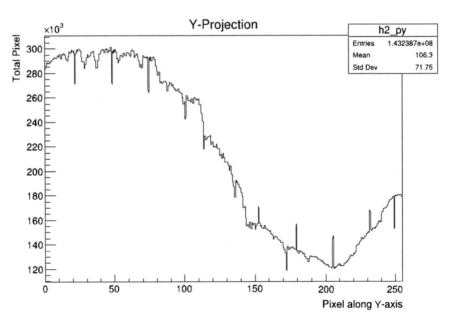

Fig. A.34 Y-projection of satellite image

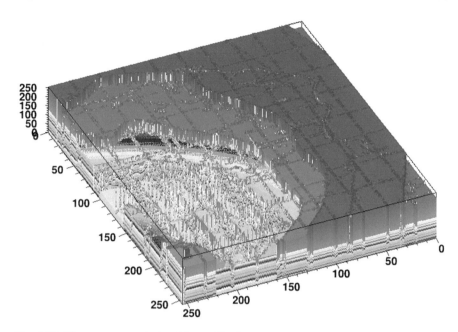

Fig. A.35 The spectrum analysis of the satellite image

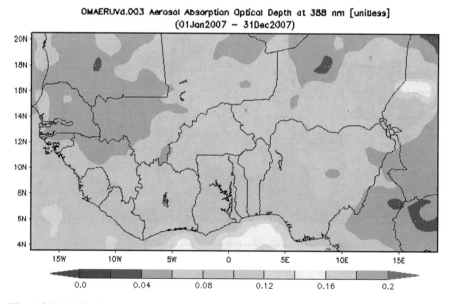

Fig. A.36 Satellite image AAOD at 388 nm, Jan. to Dec., 2007

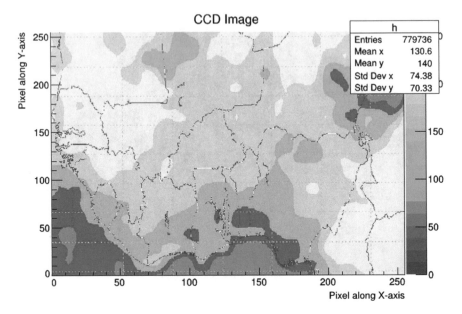

Fig. A.37 x and y pixel redefinition of the satellite image

Fig. A.38 Contour detection of satellite image

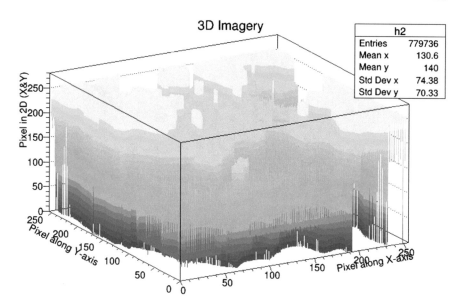

Fig. A.39 3D setting of satellite image

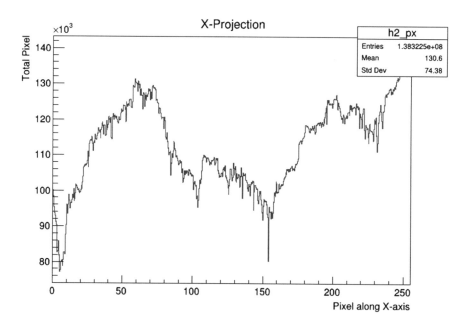

Fig. A.40 X-projection of satellite image

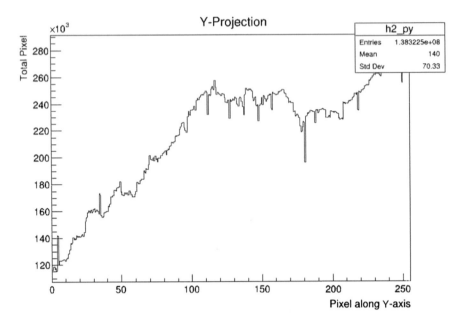

Fig. A.41 Y-projection of satellite image

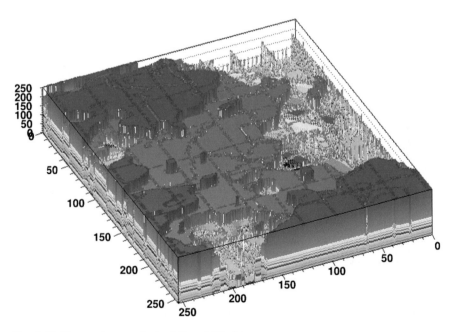

Fig. A.42 The spectrum analysis of the satellite image

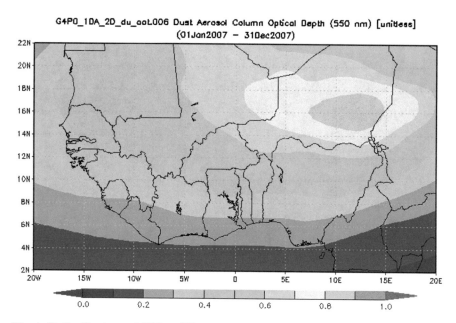

Fig. A.43 Satellite image ACOD at 550 nm, Jan. to Dec., 2007

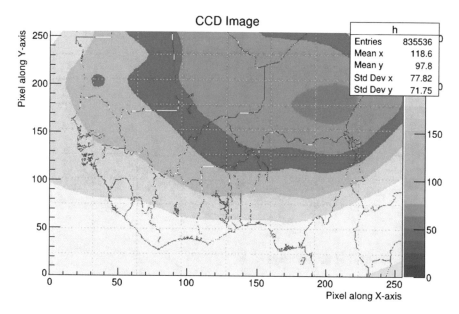

Fig. A.44 x and y pixel redefinition of the satellite image

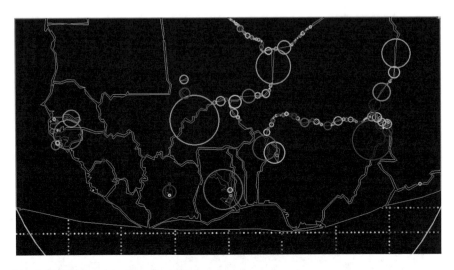

Fig. A.45 Contour detection of satellite image

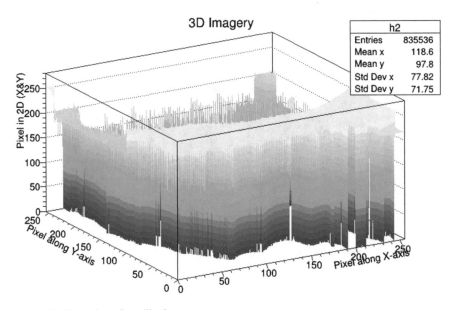

Fig. A.46 3D setting of satellite image

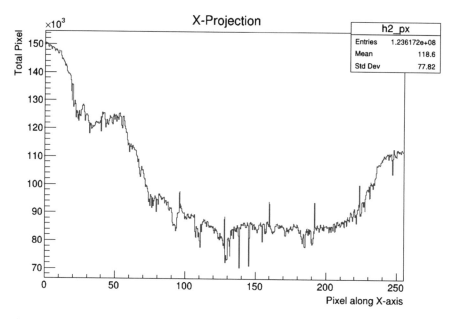

Fig. A.47 X-projection of satellite image

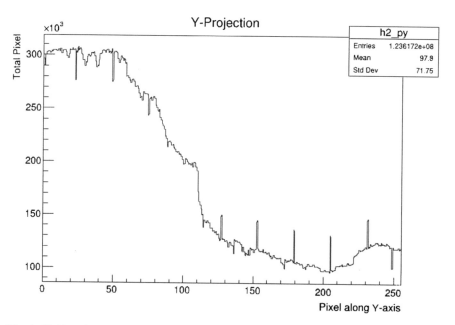

Fig. A.48 X-projection of satellite image

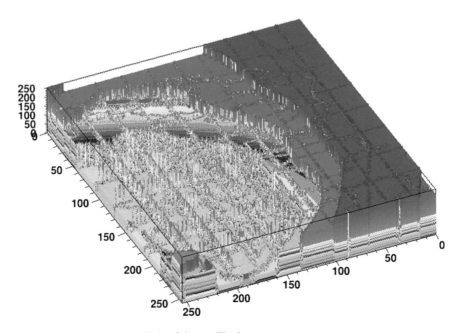

Fig. A.49 The spectrum analysis of the satellite image

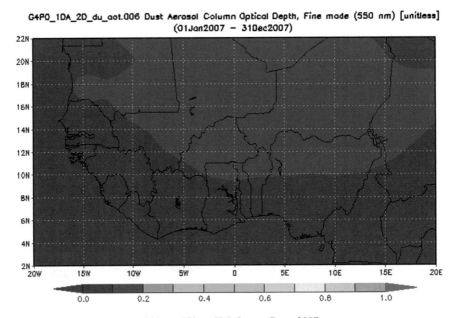

Fig. A.50 Satellite image ACOD at 550nm-FM, Jan. to Dec., 2007

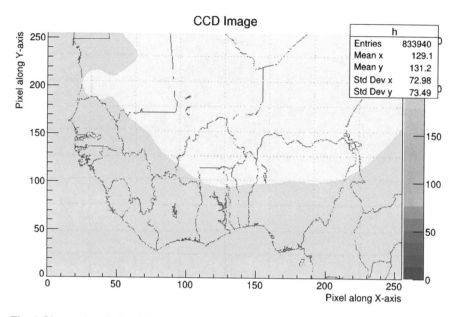

Fig. A.51 x and y pixel redefinition of the satellite image

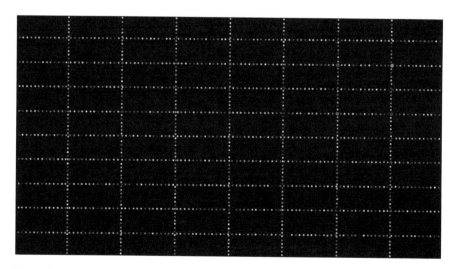

Fig. A.52 Contour detection of satellite image

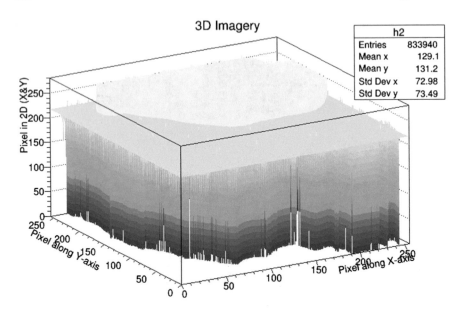

Fig. A.53 3D setting of satellite image

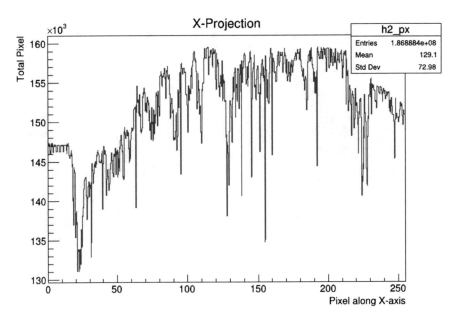

Fig. A.54 X-projection of satellite image

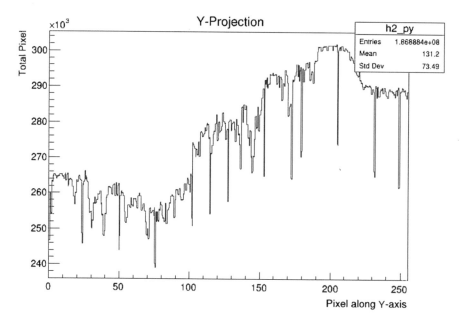

Fig. A.55 Y-projection of satellite image

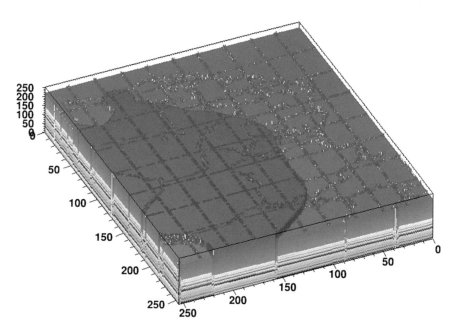

Fig. A.56 The spectrum analysis of the satellite image

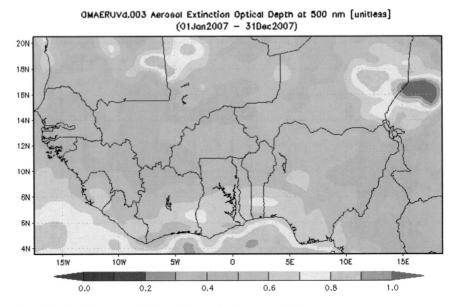

Fig. A.57 Satellite image AEOD at 500nm, Jan. to Dec., 2007

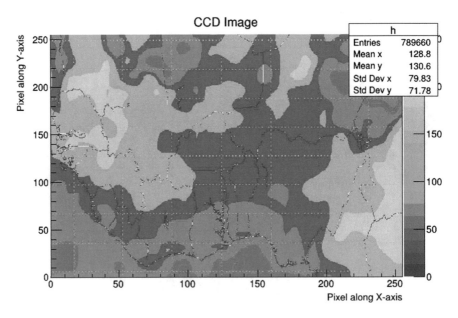

Fig. A.58 x and y pixel redefinition of the satellite image

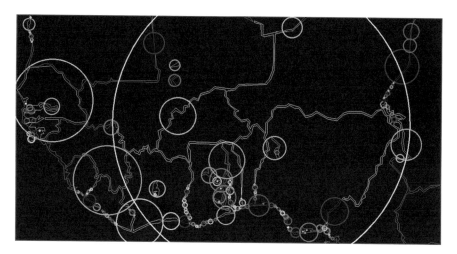

Fig. A.59 Contour detection of satellite image

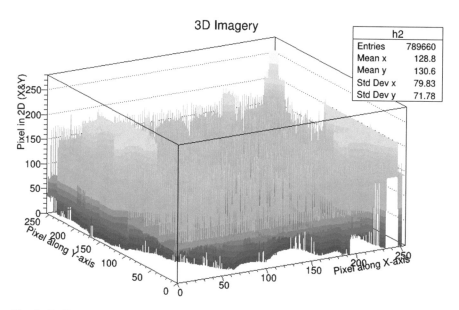

Fig. A.60 3D setting of satellite image

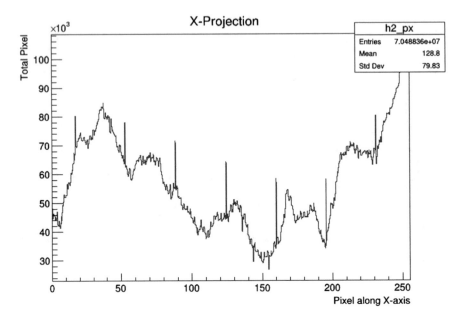

Fig. A.61 X-projection of satellite image

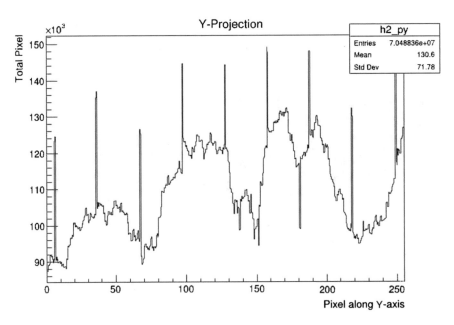

Fig. A.62 Y-projection of satellite image

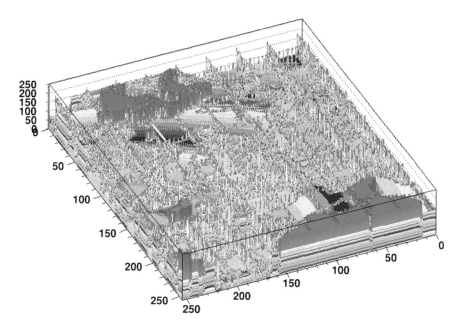

Fig. A.63 The spectrum analysis of the satellite image

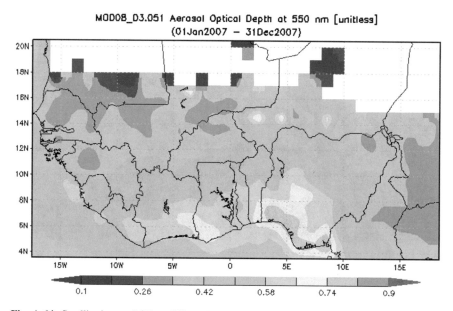

Fig. A.64 Satellite image AOD at 550nm, Jan. to Dec., 2007

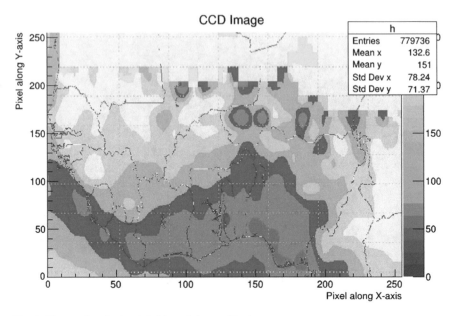

Fig. A.65 x and y pixel redefinition of the satellite image

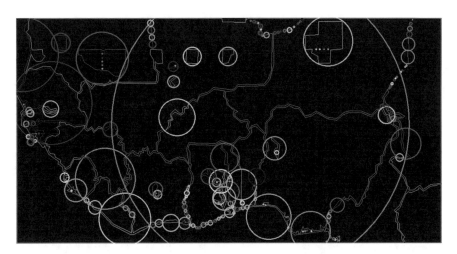

Fig. A.66 Contour detection of satellite image

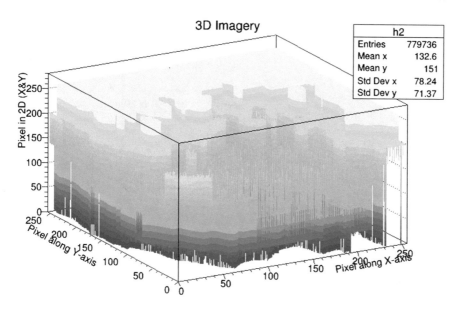

Fig. A.67 3D setting of satellite image

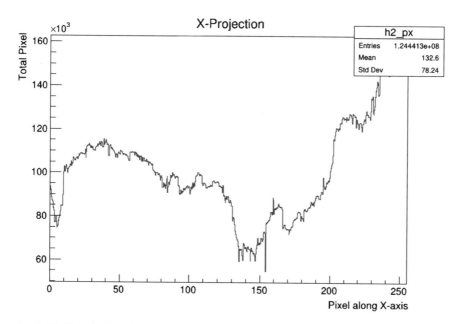

Fig. A.68 X-projection of satellite image

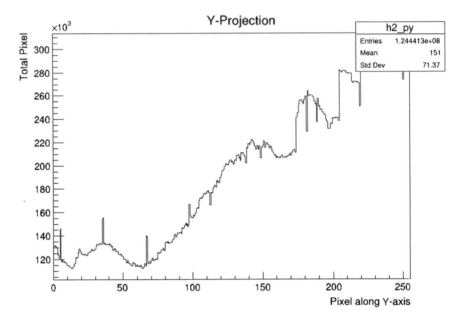

Fig. A.69 X-projection of satellite image

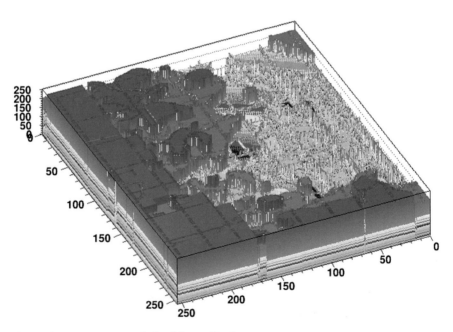

Fig. A.70 The spectrum analysis of the satellite image

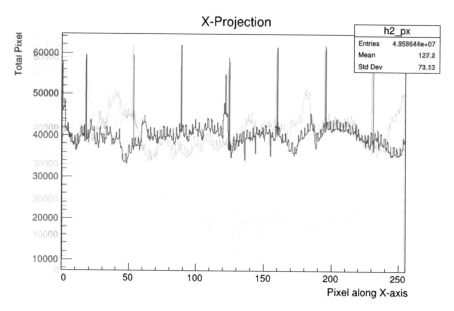

Fig. A.71 Comparative analysis of AEOD at 388 nm (2007 and 2013) and assumed standard (ACOD 2007) in the X-projection

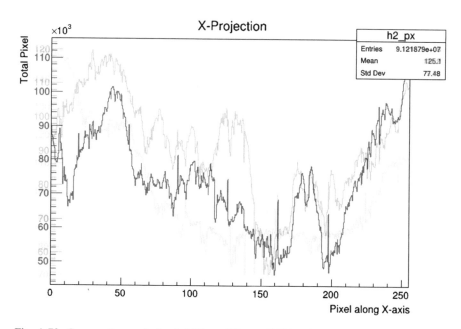

Fig. A.72 Comparative analysis of AOD at 555 nm (2007 and 2013) and assumed standard (ACOD 2007) in the X-projection

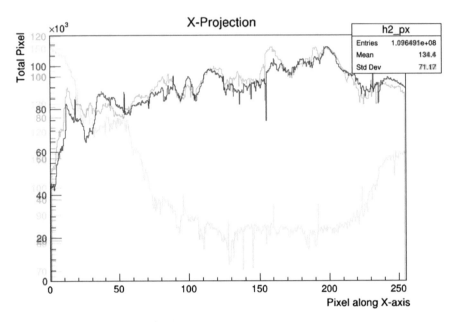

Fig. A.73 Comparative analysis of AOD pixel count (2007 and 2013) and assumed standard (ACOD 2007) in the X-projection

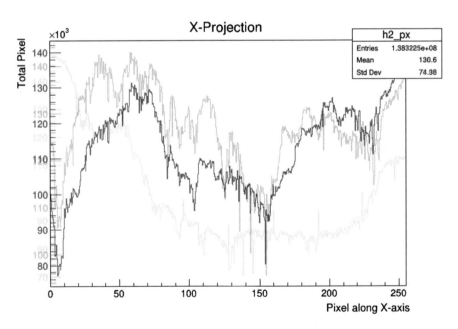

Fig. A.74 Comparative analysis of AAOD at 388 nm (2007 and 2013) and assumed standard (ACOD 2007) in the X-projection

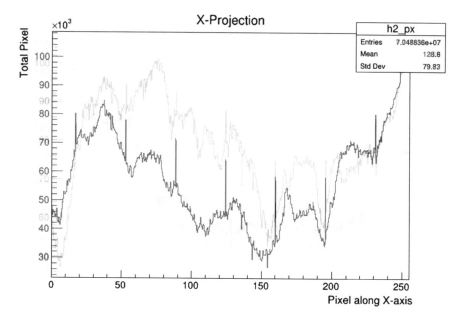

Fig. A.75 Comparative analysis of AEOD at 500 nm (2007 and 2013) and assumed standard (ACOD 2007) in the X-projection

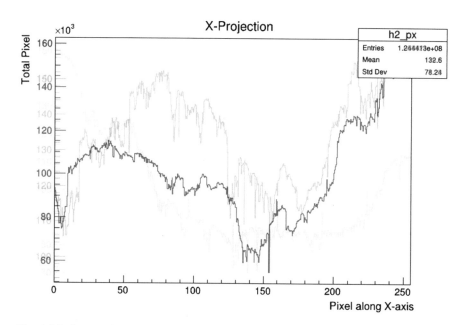

Fig. A.76 Comparative analysis of AOD at 550 nm (2007 and 2013) and assumed standard (ACOD 2007) in the X-projection

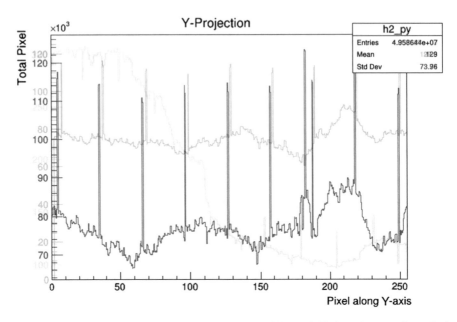

Fig. A.77 Comparative analysis of AEOD at 388 nm (2007 and 2013) and assumed standard (ACOD 2007) in the Y-projection

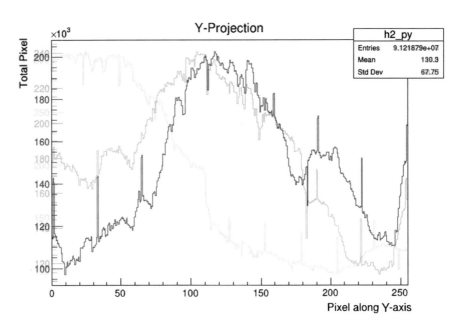

Fig. A.78 Comparative analysis of AOD at 555 nm (2007 and 2013) and assumed standard (ACOD 2007) in the Y-projection

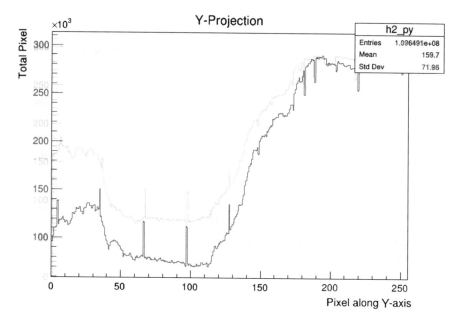

Fig. A.79 Comparative analysis of AOD pixel count (2007 and 2013) and assumed standard (ACOD 2007) in the Y-projection

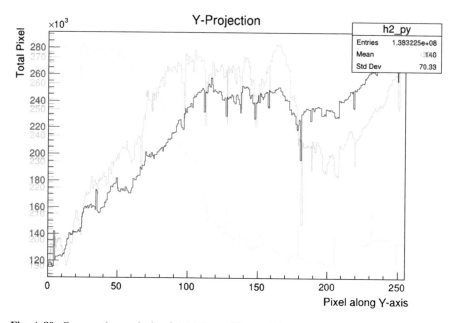

Fig. A.80 Comparative analysis of AAOD at 388 nm (2007 and 2013) and assumed standard (ACOD 2007) in the Y-projection

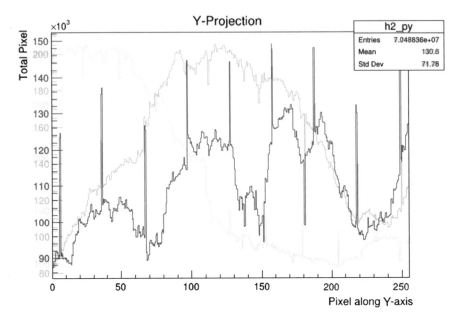

Fig. A.81 Comparative analysis of AEOD at 500 nm (2007 and 2013) and assumed standard (ACOD 2007) in the Y-projection

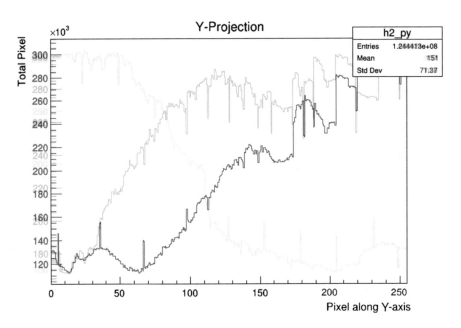

Fig. A.82 Comparative analysis of AOD at 550 nm (2007 and 2013) and assumed standard (ACOD 2007) in the Y-projection

Appendix B

See Figs. B.1 and B.2.

M. E. Emetere, *Environmental Modeling Using Satellite Imaging and Dataset Re-processing*, Studies in Big Data 54,
https://doi.org/10.1007/978-3-030-13405-1

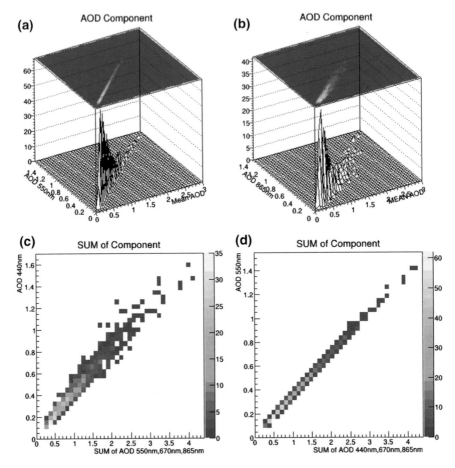

Fig. B.1 **a–d** Aerosol inter-relationality I (Chad Mao), **e–h** Aerosol daily performance I (Chad Mao), **i–l** Aerosol inter-relationality II (Chad Mao), **m–p** Aerosol inter-relationality III (Chad Mao), **q** Virtual performance of individual AOD (Chad Mao)

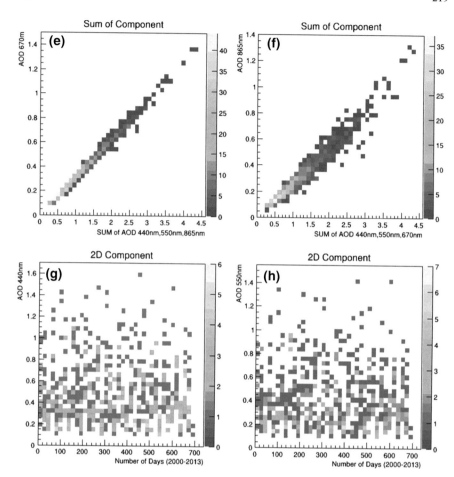

Fig. B.1 (continued)

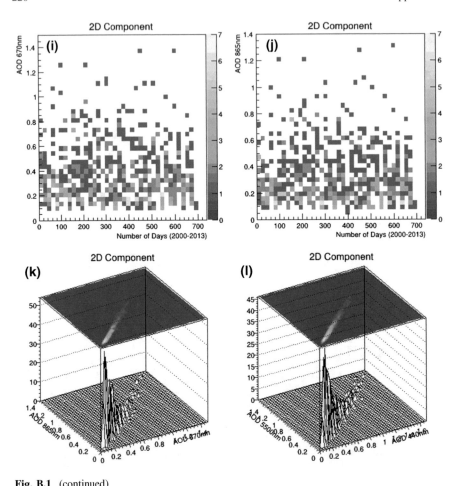

Fig. B.1 (continued)

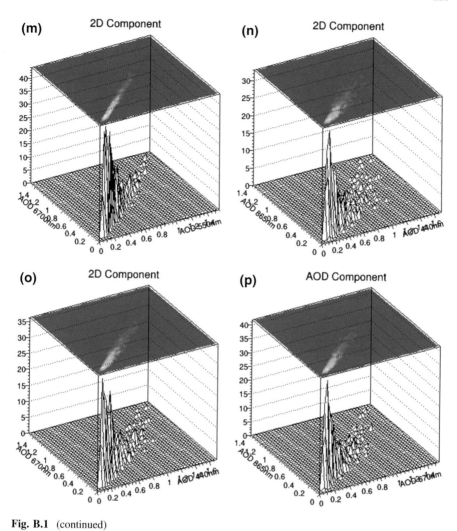

Fig. B.1 (continued)

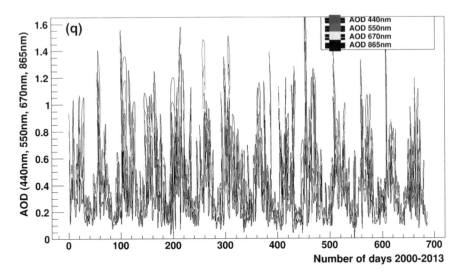

Fig. B.1 (continued)

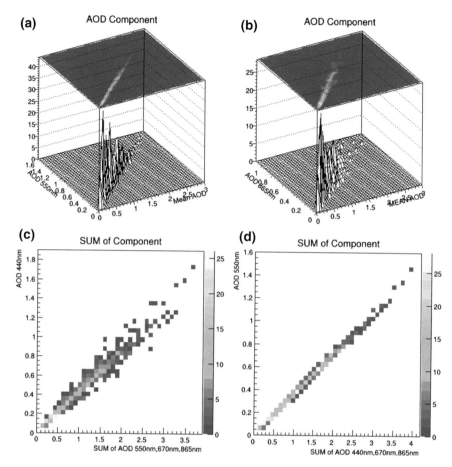

Fig. B.2 a–d Aerosol inter-relationality I (Agadez-Niger), **e–h** Aerosol daily performance I (Agadez-Niger), **i–l** Aerosol inter-relationality II (Agadez-Niger), **m–p** Aerosol inter-relationality III (Agadez-Niger), **q** Virtual performance of individual AOD (Agadez-Niger)

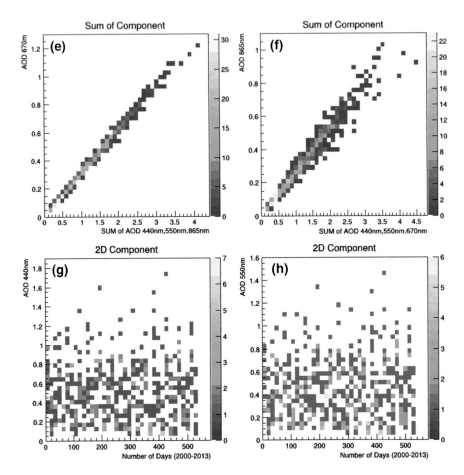

Fig. B.2 (continued)

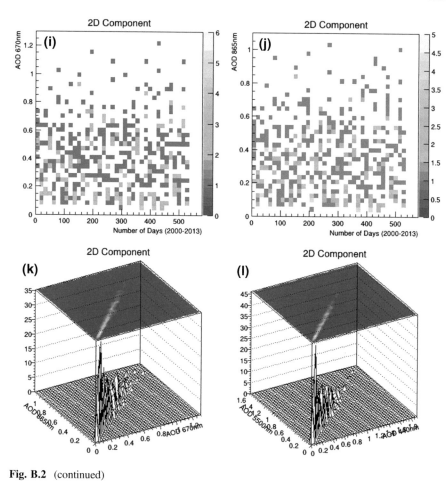

Fig. B.2 (continued)

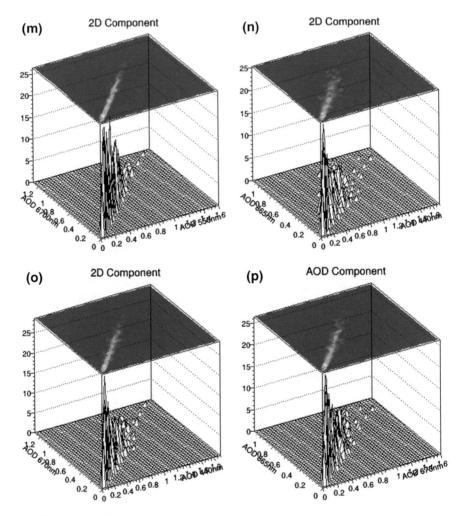

Fig. B.2 (continued)

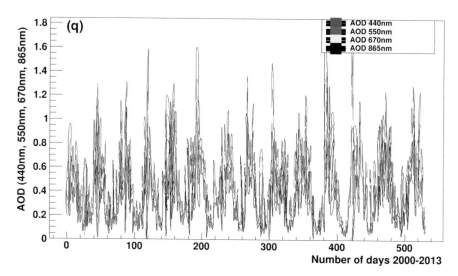

Fig. B.2 (continued)

Printed in the United States
By Bookmasters